A SYSTEMS APPROACH TO CONSERVATION TILLAGE

A SYSTEMS APPROACH TO CONSERVATION TILLAGE

Edited by Frank M. D'Itri

Institute of Water Research
 and
Department of Fisheries and Wildlife
Michigan State University
East Lansing, Michigan 48824

 LEWIS PUBLISHERS, INC.

Library of Congress Cataloging in Publication Data
Main entry under title:

A Systems approach to conservation tillage.

    Includes bibliographies and index.
    1. Conservation tillage. 2. Conservation tillage—
United States. I. D'Itri, Frank M.
S604.S97   1985   631.5′1   85-12855
ISBN 0-87371-024-X

LEWIS PUBLISHERS, INC.
121 South Main Street, P.O. Drawer 519, Chelsea, Michigan 48118

PRINTED IN THE UNITED STATES OF AMERICA

PREFACE

As the total acreage under conservation tillage has increased greatly during the past decade, its success has depended not only on the geographical location but also on the farmers' ability to match the system with crops, soil types and other environmental factors. While some farmers and environmentalists view conservation tillage as a cure-all, others see conflicts with respect to what it can accomplish relative to soil and water protection, herbicide and pesticide usage, crop yields and farm income.

Besides providing information on education, research, economic incentives, and regulations and penalties in developing strategies to encourage conservation tillage, *A Systems Approach to Conservation Tillage*[1] presents the current state of the art on methods. Besides serving as a reference document in the agronomic, horticultural, silvicultural, environmental, and agricultural engineering sciences, this book can be helpful for farmers interested in the many facets of conservation tillage.

The information ranges from scientific research projects through the experiences of farmers to educational extension programs that have extended the acceptance of conservation tillage and identified how it can protect soil and water resources while meeting human needs. The positive and negative impacts of conservation tillage technology are considered not only for plant-soil-nutrient relationships but also with respect to protecting the environment.

Consideration is also given to the ways that conservation tillage cuts production costs, reduces total labor needs, trims peak labor loads, reduces fuel requirements and provides effective wind and water erosion control. In addition, information is presented on the use of herbicides to control weeds and their increasingly complex and controversial effects on the environment.

A Systems Approach to Conservation Tillage is divided into four parts. Chapters 1 through 8 present information on the historical aspects of conservation tillage that concern the development of specialized equipment, crop management and soil systems. Chapters 9

---

[1]Michigan State University Cooperative Extension Service Publication Number NRM21.

v

through 12 describe the methods used to control weeds, insects and plant diseases. Chapters 13 through 15 relate the economics and energy requirements for selected crop systems; and Chapters 16 through 25 address environmental, public policy and sociological considerations that influence the adoption of conservation tillage practices in the United States.

This book describes concepts, value judgments and background information on the expanding conservation tillage practices in the United States and provides a technical appraisal of the state of the art. Still, much remains to be learned about the agronomic, agricultural engineering and environmental parameters; and it is hoped that the information herein presented will stimulate further research toward a more integrated approach to conservation tillage practices.

Frank M. D'Itri
East Lansing, Michigan

FRANK M. D'ITRI

Frank M. D'Itri is a Professor of Water Chemistry in the Institute of Water Research and Department of Fisheries and Wildlife at Michigan State University in East Lansing, Michigan. He holds a Ph.D. in analytical chemistry with special emphasis on the transformation and translocation of phosphorus, nitrogen, heavy metals and hazardous organic chemicals in the environment.

Professor D'Itri is listed in *American Men of Science, Physical and Biological Science*. He is the author of *The Environmental Mercury Problem*, co-author of *Mercury Contamination* and *An Assessment of Mercury in the Environment*. He is the editor of *Wastewater Renovation and Reuse, Land Treatment of Municipal Wastewater: Vegetation Selection and Management, Municipal Wastewater in Agriculture, Acid Precipitation: Effects on Ecological Systems, Artificial Reefs: Marine and Freshwater Applications* and co-editor of *PCBs: Human and Environmental Hazards* and *Coastal Wetlands*. In addition, he is the author of more than forty scientific articles on a variety of environmental topics.

Dr. D'Itri has served as chairperson for the National Research Council's panel reviewing the environmental effects of mercury and has conducted numerous symposia on the analytical problems related to environmental pollution. He has been the recipient of fellowships from Socony-Mobil, the National Institutes of Health, the Rockefeller Foundation and the Japan Society for the Promotion of Science.

To the

COOPERATIVE EXTENSION SERVICE

and the

AGRICULTURAL EXPERIMENT STATION

Michigan State University
East Lansing, Michigan 48824

ACKNOWLEDGMENTS

The success of this endeavor was due to the interest and dedication of the participating individuals and organizations. Primary among these was the Michigan State University College of Agriculture and Natural Resources, Cooperative Extension Service and Agricultural Research Station which sponsored and contributed to the necessary financial support for the conference and publication of this book. My special thanks are to Dr. Adger B. Carroll, Assistant Director for Natural Resources and Public Policy, MSU Cooperative Extension Service; Dr. Robert G. Gast, Director, MSU Agricultural Experiment Station and Mr. Homer R. Hilner, State Conservationist, USDA Soil Conservation Service, for their encouragement and support during the preliminary planning as well as their direct participation in the conference itself. I would also like to thank Dr. Harvey J. Liss, Extension Program Leader, and Ms. Patsy Kreuzer, along with the many other staff members of the Kellogg Biological Station Conference Center, not only for their hospitality and assistance but also for providing the conferees with an atmosphere that was free from distractions and conducive to the discussion of the many diverse aspects related to conservation tillage. Last, but certainly not least, a grateful acknowledgment also goes to all of the authors whose research, in addition to contributions of time, efforts, and counsel made this volume possible.

In the Institute of Water Research at Michigan State University, special thanks are extended to its Director, Dr. Jon F. Bartholic, for his support and encouragement. In the preparation of this manuscript, I acknowledge and thank Dr. Charles S. Annett and Ms. Dara J. Philipsen for many editorial suggestions that contributed materially to the overall style.

A SYSTEMS APPROACH TO CONSERVATION TILLAGE
CONTRIBUTORS

George W. Bailey, U.S. Environmental Protection Agency, Environmental Research Laboratory, College Station Road, Athens, GA 30613

James L. Baker, Professor, Department of Agricultural Engineering, Iowa State University, Ames, IA 50011

David B. Beasley, Associate Professor, Department of Agricultural Engineering, Purdue University, West Lafayette, IN 47907

Louis B. Best, Professor, Department of Animal Ecology, Iowa State University, Ames, IA 50011

Ernest E. Behn, Route 1, Boone, IA 50036

J. Roy Black, Department of Agricultural Economics, Michigan State University, East Lansing, MI 48824

Robert L. Blevins, Professor, Agronomy Department, University of Kentucky, Lexington, KY 50546

Donald Christenson, Professor, Department of Crop and Soil Sciences, Michigan State University, East Lansing, MI 48824

Larry J. Conner, Department of Agricultural Economics, Michigan State University, East Lansing, MI 48824

William H. Darlington, former Graduate Student, Department of Crop and Soil Sciences, Michigan State University, East Lansing, MI 48824

Frank M. D'Itri, Institute of Water Research, Michigan State University, East Lansing, MI 48824

Boyd G. Ellis, Professor, Department of Crop and Soil Sciences, Michigan State University, East Lansing, MI 48824

W. W. Frye, Professor, Agronomy Department, University of Kentucky, Lexington, KY 40546

Arthur J. Gold, Assistant Professor, University of Rhode Island, Kingston, RI 02881

Anthony M. Grano, Economic Research Service, U.S. Department of Agriculture, GHI Building, Room 428, 500 12th Street S.W., Washington, DC 20250

Donald R. Griffith, Extension Agronomist, Agronomy Department, Purdue University, West Lafayette, IN 47907

William A. Hayes, Sales Manager, Industrial Site, Box 848, Fleischer Manufacturing, Inc., Columbus, NE 68601

Homer R. Hilner, State Conservationist, Soil Conservation Service, U.S. Department of Agriculture, 101 Manly Miles Building, 1405 South Harrison Road, East Lansing, MI 48823

Maureen K. Hinkle, Agricultural Policy Coordinator, National Audubon Society, 645 Pennsylvania Avenue S.E., Washington, DC 20003

James J. Kells, Assistant Professor, Department of Crop and Soil Sciences, Michigan State University, East Lansing, MI 48824

H. Walker Kirby, Assistant Professor and Extension Specialist, Department of Plant Pathology, University of Illinois at Urbana-Champaign, N531 Turner Hall, 1102 South Goodwin Avenue, Urbana, IL 61801

James E. Lake, Field Office Coordinator, Conservation Tillage Information Center (CTIC), Executive Park, 2010 Inwood Drive, Fort Wayne, IN 46815

Lawrence W. Libby, Department of Agricultural Economics, Michigan State University, East Lansing, MI 48824

Ted Loudon, Associate Professor, Department of Agricultural Engineering, Michigan State University, East Lansing, MI 48824

Jerry V. Mannering, Professor, Department of Agronomy, Purdue University, West Lafayette, IN 47907

W. F. Meggitt, Professor, Department of Crop and Soil Sciences, Michigan State University, East Lansing, MI 48824

William C. Moldenhauer, Soil Scientist, National Soil Erosion Laboratory, Agricultural Research Service, U.S. Department of Agriculture, West Lafayette, IN 47907

John C. Moore, Natural Resources Ecology Laboratory, Colorado State University, Fort Collins, CO 80523

H. Muhtar, Department of Agricultural Economics, Michigan State University, East Lansing, MI 48824

Lee A. Mulkey, U.S. Environmental Protection Agency, Environmental Research Laboratory, College Station Road, Athens, GA 30613

Peter J. Nowak, Associate Professor, University of Wisconsin-Madison, Environmental Resource Unit, 216 Agricultural Hall, Madison, WI 53706

Francis J. Pierce, Department of Crop and Soil Sciences, Michigan State University, East Lansing, MI 48824

J. Posselius, Department of Agricultural Economics, Michigan State University, East Lansing, MI 48824

Dwight Quisenberry, Soil Conservation Service, U.S. Department of Agriculture, 101 Manly Miles Building, 1405 South Harrison Road, East Lansing, MI 48823

Charles W. Rice, Department of Crop and Soil Sciences, Michigan State University, East Lansing, MI 48824

C. Alan Rotz, Agricultural Engineer, Agricultural Research Service, U.S. Department and Agriculture and Department of Agricultural Engineering, Michigan State University, East Lansing, MI 48824

Robert F. Ruppel, Department of Entomology, Michigan State University, East Lansing, MI 48824

Gerald D. Schwab, Department of Agricultural Economics, Michigan State University, East Lansing, MI 48824

Kathleen M. Sharpe, Department of Entomology, Michigan State University, East Lansing, MI 48824

M. S. Smith, Department of Agronomy, University of Kentucky, Lexington, KY 50546

Richard J. Snider, Department of Zoology, Michigan State University, East Lansing, MI 48824

Jusup Subagja, Faculty of Biology, Gadjah Mada University, Yogyakarta, Indonesia

Robert R. Swank, Jr., U.S. Environmental Protection Agency, Environmental Research Laboratory, College Station Road, Athens, GA 30613

Glover B. Triplett, Jr., Professor, Agronomy Department, Box 5248, Mississippi State University, Mississippi State, MS 39762

David M. Van Doren, Jr., Professor, Department of Agronomy, Ohio Agricultural Research and Development Center, and Ohio State University, Wooster, OH 44691

Maurice L. Vitosh, Professor, Department of Crop and Soil Sciences, Michigan State University, East Lansing, MI 48824

A SYSTEMS APPROACH TO CONSERVATION TILLAGE
F.M. D'Itri, Editor

CONTENTS

PART I

HISTORICAL, EQUIPMENT DEVELOPMENT, CROP
MANAGEMENT, AND SOIL SYSTEMS

XV

PART II

CONTROL OF WEEDS, INSECTS, AND PLANT DISEASES

PART III

THE ECONOMICS AND ENERGY REQUIREMENTS FOR
SELECTED CONSERVATION TILLAGE CROP SYSTEMS

PART IV

CONSERVATION TILLAGE: ENVIRONMENTAL, PUBLIC POLICY,
AND SOCIOLOGICAL CONSIDERATIONS

PART I

HISTORICAL, EQUIPMENT DEVELOPMENT, CROP MANAGEMENT, AND SOIL SYSTEMS

CHAPTER 1

A SYSTEMS APPROACH TO CONSERVATION TILLAGE: INTRODUCTION

F. J. Pierce
Crop and Soil Sciences Department
Michigan State University
East Lansing, Michigan

INTRODUCTION

Conservation tillage has come of age. This is reflected in the significant rate of adoption of conservation tillage characteristic of the 1980s (Quisenberry, Chapter 7), in its prominent role in public policy related to conservation issues in the United States, and in its historical context (its roots extend to the 1940s). Its growth has necessitated the establishment of the Conservation Tillage Information Center (CTIC) in 1983 with the purpose of collecting and disseminating information on conservation tillage farming techniques (Lake, Chapter 26). Problems associated with the aging process are clearly expressed in the uncertainties surrounding the physical, chemical, and biological impacts of cultural practices characteristic of conservation tillage (Baker, Chapter 18; Best, Chapter 23; Ellis et al., Chapter 21; Hinkle, Chapter 22; Snider et al., Chapter 14). Questions of economics, both in the short term and the long term, and the sustainability of agriculture are important in the analysis of conservation tillage systems.

Conservation tillage is a complex, multi-faceted subject. Therefore, in considering conservation tillage both now and in the future, one must look at these many facets in a systems context. This conference adopted a theme of a systems approach to conservation tillage. My assignment is to introduce the proceedings of this conference and I will do so by discussing the many facets of conservation tillage and their interactions in the context of the papers that follow.

Definition

A classic problem affiliated with conservation tillage over its years of development has been its definition. Cultural practices associated with crop production vary considerably with region of the country such that confusing terminology emerged. At least five of the papers in this proceedings provide some definition of conservation tillage. Crosson (1981) stated that there is no precise commonly accepted definition of conservation tillage. Mannering and Fenster (1983) suggested that conservation tillage has a broad definition but requires the reduction of soil and water losses relative to conventional tillage. At present, the Soil Conservation Service (SCS) defines conservation tillage as any system which leaves 30 percent residue cover after planting. More recent research suggests the value of tillage-induced roughness and type and incorporation of crop residue in reducing soil loss reductions (Cogo et al., 1984) may need to be included in any operational definition. It may be appropriate for these proceedings, therefore, to accept Mannering and Fenster's (1983) suggestion that all tillage practices that conserve soil and water be listed under the term conservation tillage, keeping in mind that much of the discussion in the following papers deals with tillage systems where crop residues are managed at or near the soil surface.

Soil and Water Conservation

The basic impetus for research on conservation tillage came from the need to control soil erosion and conserve soil moisture. It is clear that management of crop residues on or near the surface provides considerable erosion control and moisture conservation (see Hayes, Chapter 3; Blevins et al., Chapter 9; and Moldenhauer, Chapter 10). Different combinations of tillage tools and different crops will result in markedly different amounts of crop residues on the soil surface. It is important, therefore, that consideration be given to the amount of residue produced by the previous crop. Care must be taken when selecting tillage tools and when performing tillage operations to insure adequate residue cover for erosion protection and water conservation. Peter Nowak (Chapter 24) demonstrated that farmers' success at achieving conservation objectives with crop residue management may be significantly less than indicated by adoption statistics being reported. Technology transfer and education of farmers in conservation

methods needs more emphasis to increase successful adoption
of conservation tillage systems.

Crop Production

There are two essential components of crop production
systems: they must be sustainable and they must be profitable.
Sustainable agricultural requires the maintenance of long-term
soil productivity, and therefore, requires the protection
of the soil resource base.   It also requires increased
production per unit land area and the protection of water
resources, both surface and ground water, to meet future
food and fiber needs asssociated with world population
growth.   Profitable agriculture requires net farm growth
realized through proper combinations of high yields, reduced
production costs, good commodity prices, and low interest
rates.   The first two are addressable at the farm level,
the latter two have eluded the farmer's control for the
most part.   Sustainability and profitability have often
been at odds over the history of soil conservation in
the United States. The costs of many traditional conservation
practices to farmers may be greater, in the short run,
then the return from controlling erosion in terms of sustaining
long term productivity (Rosenberry et al., 1981).   It
appears, however, in the 1980s, that conservation tillage
may have brought these two components closer together
than at any time in history.   In fact, many farmers have
successfully adopted conservation tillage systems resulting
in harmony between these two fundamental components.
In fact, economic incentives of conservation tillage systems
are a major reason for adoption of conservation tillage
systems (Nowak, Chapter 24; Conservation Tillage Information
Center, 1984).

Successful crop production utilizing conservation
tillage techniques requires technology, knowledge of it
and skill in its use. Nearly every aspect of crop production--
planting; equipment performance; fertilizer practices;
weed, insects, and disease incidence and control--are
affected in some way by the reduction in tillage and presence
of crop residues. Each component of the resource management
system must be understood and properly tuned for the tillage
system to produce the desired result--maximum profit margin.

EQUIPMENT

Many of the conservation tillage tools are adaptations
of existing equipment.   Disks, both offset and tandem
disks, were adopted for conservation tillage without much
change.   Some tools went through significant modification.

A good example of this is the chisel plow, which was first modified to provide better clearance to avoid clogging with residues. Later, a row of coulter blades was added in front of the chisels to cut up crop residue, thereby incorporating residue disking and chiseling into one operation. The coulter or disk chisel is a common conservation tillage tool and has become the conventional system in many areas replacing the moldboard plow. Considerable development has taken place in field cultivators, especially in regards to herbicide incorporation with residues present.

Perhaps the major advance in equipment has been in planters, both in row crop planters and grain drills. Basically, planters were modified to increase penetration in the soil and to cut through residues. This was accomplished by utilizing various coulters or disk openers in combination with increased weight and various methods of increasing down pressure. Other modifications were made to allow additional cultural practices, such as herbicide and fertilizer applications, to be combined with planting. Specialized equipment has been developed, such as ridge till and till-plant cultivators and planters (Hayes, Chapter 3 and Behn, Chapter 4), or in-row subsoilers, or tools to deep till without disturbance of surface residue cover; all for the purpose of developing profitable and sustainable agricultural systems. Farmer innovation, in particular, in conservation tillage has led to many changes in equipment design. Basically, the key to conservation tillage equipment is this: there are many manufacturers of implements and planters that have perform well in conservation tillage systems and will prove equally satisfactory for a specific task. Most important is the appropriate selection of equipment (Rotz and Black, Chapter 16) and its proper adjustment to attain the desired outcome; whether that be seed placement, tillage depth or residue cover, or fertilizer placement (Hayes, Chapter 3).

FERTILIZER

The behavior of fertilizer, especially nitrogen, as affected by conservation tillage systems has received considerable research effort. Vitosh **et al.** (Chapter 8) provide a brief overview of the pertinent research and main points of interest with regards to fertilizer management. Conservation tillage significantly affects the behavior and distribution of essential nutrients in the soil and the chemical condition of the soil, especially as expressed by soil pH. The chemical as well as the microbiological and physical effects of conservation tillage on soil properties is amply discussed by Blevins **et al.** (Chapter 9). The most current information available, however, indicates that through soil testing and proper fertilizer and lime management, fertility will not be a limiting factor for high yield crop production with conservation tillage systems in Michigan.

CROP YIELD

Not all soils are adapted to conservation tillage systems given present available technology. The state of available technology is an important consideration in the discussion of conservation tillage systems, since the rapid adoption of conservation tillage in the United States in the last ten years is to a large extent due to technological developments in equipment and herbicides. One must ask the question: "What will the impact of the biotechnological revolution now in progress have on agriculture by the turn of the century?" It is also well to remember the innovation of the farmer, who traditionally takes on the challenge of "you can't make that work!"

Griffith and Mannering (Chapter 5) and Triplett and Van Doren (Chapter 6) present data showing the capacity of conservation tillage systems for corn and soybean production. Factors important in determining successful crop production with conservation tillage include drainage, the previous crop (rotations), climate, operator skill, fertilizer practices, soil condition (compaction), and pest problems. It is generally agreed, as stated in both papers, that as drainage improves, the need for tillage decreases. Multiple cropping of wheat-soybeans and silage corn following hay harvest are among potential benefits of conservation tillage systems where such management practices are possible (Triplett and Van Doren, Chapter 6).

For states such as Michigan, where agriculture is diverse (many different crops and considerable dairy, livestock and feeding operations), the role of conservation tillage systems is not yet fully understood, since much of the conservation tillage research has focused on corn and soybean production. Early planting advantages are associated with direct drilled spring grains. Conservation aspects of no-till forage production on erosive soils, wind erosion protection on sandy and organic soils (Michigan ranks fifth in the nation for sandy cultivated lands and second for organic soils), cover crops for silage corn, and benefits of legumes on succeeding corn crops as affected by tillage are some of the many topics on the research agenda for the future.

WEED CONTROL

There is more interest in and controversy over weed control and herbicide use relative to conservation tillage than any other tillage-related topic. Greater reliance on herbicides for weed control is a predominant characteristic of conservation tillage systems. In fact, lack of adoption of conservation tillage systems for some crops, such as dry beans, may in part be due to the lack of weed control options.

Kells and Meggitt (Chapter 11) provide a concise overview of conservation tillage-weed control interactions. It is clear that weed control strategies must change with tillage systems to accommodate changes in weed behavior, weed spectrum, weed pressures, as well as herbicide performance associated with changes in soil properties (Blevins et al., Chapter 9) and the presence of crop residues on the soil surface. They suggest that effective weed control in conservation tillage systems is possible when herbicides are properly managed. Knowing your weeds and chemicals and keeping up to date with university recommendations will lead to good weed control.

INSECTS AND PLANT DISEASES

Kirby (Chapter 12) presents a succinct summary of the potential disease problems and benefits associated with conservation tillage and related cultural practices. He indicated that conservation tillage offers potential for increased biological activity associated with crop residues. Tillage induced changes in soil environment may lead to beneficial or adverse effects on either the plant or disease pathogens. High levels of protection from common plant disease problems associated with residues are attainable with proper management practices and pesticides. Management practices for disease control include modification of planting date, row spacing and plant populations as well as crop rotations, and use of disease resistant varieties. Regular scouting of fields by persons trained in Integrated Pest Management (IPM) methods will also contribute to better control.

Insects, like weeds, are of particular interest because insect spectrum and pressures also shift with changing tillage. This is largely associated with crop residues, use of cover crops such as rye, and the presence of grass weeds if control is inadequate (Rupple and Sharpe (Chapter 13). The importance of regularly scouting fields for pests is critical since insects are always most damaging when they are unexpected (Rupple and Sharpe, Chapter 13).

The position taken by Rupple and Sharpe that pest outbreaks are waiting to happen, that sooner or later some species will succeed, and as they state "we must still fear the unknown" leaves me concerned about the state of understanding of the biology of conservation tillage.   If their gut feelings are true, agriculture has a significant problem. Either way, considerable research on soil biology of conservation tillage is warranted.

ECONOMICS

Profitability in agricultural production is essential to the farmer.   Sustainability of agriculture is critical to society. The goal is harmony between the two.   That profitability takes priority is quite clear, especially in light of the farm crisis of the 1980s.   The economics of conservation tillage systems are becoming known and in many situations these systems show clear economic advantages (Rotz and Black, Chapter 16; Black et al., Chapter 17; Hayes, Chapter 3; Grano, Chapter 15; Griffith and Mannering, Chapter 5).   The rapid adoption of conservation tillage has been attributed to the existence of an economic advantage (Conservation Tillage Information Center, 1984).

The direct beneficiary of this economic incentive should be soil and water conservation. However, conservation practices are frequently not adequate for erosion control (Moldenhauer, Chapter 10; Nowak, Chapter 24) and are not being adopted on the more erosive cropland.   Figure 1 shows a plot of the average potential for soil erosion (RKLS product) versus the C and P factors of the Universal Soil Loss Equation (USLE) for row and close grow crops grown nationally in the United States in 1982 (Pierce et al., 1985).   Data are weighted averages for row and close grown crops summarized nationally from data reported in the 1982 National Resources Inventory.   Both the P and C factors should decrease as the potential for soil erosion conservation practices were being applied to the more erosive land.   Since this is not the case, it must be concluded that erosion control is not the major contributing factor in the adoption process.

The major conclusion from economic studies thus far is that conservation tillage systems for which crop yields are near or equal to those obtained conventionally have shown an economic advantage. However, many factors contribute to the determination of the cost/benefit ratio of various tillage systems. This very important calculation, therefore, should be performed for each individual situation. Since assumptions made and calculation methods employed

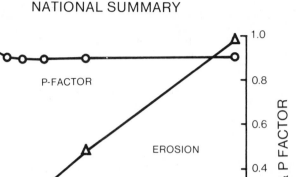

Figure 1. Plots of the 1982 NRI weighted average erosion rate (t/ac/yr), C-factor and P-factor versus the potential for erosion (t/ac/yr) as expressed by the RKLS product of the USLE. Data is summarized nationally. (Pierce et al., 1985)

vary, different models will likely lead to different outcomes. It is important in the near future for economists to develop realistic models and make them available to farmers enabling them to make informed decisions about resource management systems they may choose to adopt.

PUBLIC POLICY

Conservation tillage has become a formidable tool for the Soil Conservation Service in their work to control soil erosion (Hilner, Chapter 2). The impact of conservation tillage on public policy in the United States has been monumental. Most recently, the Lansing State Journal (Sunday, June 30, 1985) reported that the Secretary of Agriculture, John Block, "proposes returning 20 million acres of erodible cropland to nature. The 10-year, $11 billion program would be the largest soil conservation initiative in United States history, he said." Libby

(Chapter 25) discusses the policy issues related to soil and water conservation--the whys, the relevance, and the alternatives for policy--who pays, and who has what rights in the matter of soil and water conservation. While it may at times be confusing, such policy matters are important and, as indicated by the news report, can be quite costly.

ENVIRONMENTAL CONCERNS

Five papers in this conference proceedings discuss the environmental and ecological concerns regarding the fate of eroded soil and applied chemicals (fertilizers and pesticides) as they relate to conservation tillage and related cultural practices. With regard to water quality considerations, Baker (Chapter 18) concluded that reduced sediment and sediment-associated chemical losses associated with conservation tillage are obvious. Less obvious, however, is the potential for increased chemical losses due to increased chemical concentrations in the runoff associated with an increase in surface-applied chemicals in conservation tillage systems. Ellis et al. (Chapter 21) present data that support Baker's conclusions. Best (Chapter 23), Hinkle (Chapter 22) and Snider et al. (Chapter 14) provide some insight into the effect of conservation tillage and chemical applications on non-target organisms, such as wildlife and soil arthropods. It is clear from these papers that much work remains with regards to the biological responses, whether beneficial or adverse, associated with of conservation tillage. Much of the environmental and ecological concerns are directly related to the use of pesticides.

INFORMATION TRANSFER

Information is a critical ingredient for the successful adoption and use of conservation tillage systems. Hilner (Chapter 2) expresses the need for leadership and cooperation among the agencies in Michigan through whose efforts conservation tillage systems will play a major role in sustainability and profitability of Michigan's agriculture. The role of the university and the Cooperative Extension Service is clear in terms of technology development in resource management systems and the subsequent effective transfer of that technology to the farmer. The Conservation Tillage Information Center can make a significant contribution in information collection and dissemination. The role of computers in resource management and decision making is developing (Beasley, Chapter 20).

Table 1.   General questions about conservation tillage
expressed by conference participants.

TILLAGE SYSTEMS PERFORMANCE

--Effect of tillage practices on farm profit.
--Which conservation tillage practice is most effective
   in improving soil tilth?
--Can we use the same conservation tillage practices on
   all agricultural soils?
--How much additional herbicides and pesticides are needed
   for conservation tillage and why?
--Do I need to change my fertilizer practices if I change
   my tillage practices?
--Where and how can agrochemicals (fertilizers and pesticides)
   be placed in no-till and ridge till systems for maximum
   use by the crop so that they won't be lost?

ENVIRONMENTAL EFFECTS

--Effect of tillage practices on surface hydrology.
--Effect of tillage practices on major transport and transportation
   processes.
--What are the wind and water erosion impacts from the
   different conservation tillage systems?
--How effective is conservation tillage abatement on improved
   water quality?
--How can conservation tillage methods reduce the impact
   from agricultural non-point pollution?
--Can a land and water model be modified or devised so
   it accurately describes the potential exports and
   erosion from a watershed or river basin?

ECOLOGICAL CONSIDERATIONS

--Long- and short-term effects of conservation tillage
   on microbial ecology.
--What are the long-term effects of conservation tillage
   on weeds?
--Effect of tillage practice (both type and timing) on
   wildlife populations.
--Are soil animals good for soil stability?
--Will best species occur in greater frequency in reduced
   tillage and increased residues situations?

FUTURE NEEDS

Research on conservation tillage has been quite profitable, but many research problems remain. Some of the more prominant ones identified by the participants are included in Table 1. Others have been identified within individual papers in these proceedings. It is quite clear from these lists that the biological aspects of conservation tillage warrant considerable research. Significant technological advances may be developed from biotechnological research currently underway. Intriguing developments may also be expected from advances in artificial intelligence and expert systems, robotics, and decision support technology. Much of this technology may allow farmers to match equipment and cultural practices to the soils and landscapes within their fields. Whatever the research agenda, transferring new technologies to the farmer will be increasingly important.

SUMMARY

I have attempted to provide a brief overview of conservation tillage while pointing the reader to appropriate papers in these proceedings. The papers presented here cover many aspects of conservation tillage, all of which must be considered together to understand the complexity of these tillage systems. There are no magic recipies for success nor are there any guarantees included in these pages; only sound advice, lots of experience, and many ideas for your consideration.

LITERATURE CITED

Cogo, N.P., W.C. Moldenhauer, and G.R. Foster. 1984. Soil loss reductions from conservation tillage practices. Soil Sci. Soc. Am. J. 48:368-373.

Conservation Tillage Information Center. 1984. A Survey of America's Conservation Districts. Fort Wayne, IN.

Crosson, P. 1981. Conservation tillage and conventional tillage: A comparative assessment. Soil Cons. Soc. Am., Ankeny, Iowa 50021. 34 p.

Mannering, J.V. and C.R. Fenster. 1983. What is conservation tillage? J. Soil and Water Cons. 38:141-143.

Pierce, F.J., W.E. Larson, and R.H. Dowdy. 1985. Field estimates of C factors: How good are they and how do they affect calculation of erosion. A paper presented at the National Academy of Sciences Convocation on Physical

Dimensions of the Erosion Problem, December 7, 1984. (In press).

Rosenberry, P., R. Knutson, and L. Harmon. 1980. Predicting the effect of soil depletion from erosion. J. Soil and Water Cons. 35:131-134.

CHAPTER 2

LEADERSHIP IN CONSERVATION TILLAGE

Homer R. Hilner
U.S. Department of Agriculture
Soil Conservation Service
East Lansing, Michigan 48823

INTRODUCTION

Harrold Scharer won the 1983 Michigan No-Till Contest sponsored by the Ortho Chemical Company with a corn yield of over 10,350 kg/ha. That's leadership in conservation tillage! Mr. Scharer happens to farm in Saginaw County; however, there were 137 entrants in this contest from 26 Michigan counties. The average no-till corn yield was 7213 kg/ha. I can remember when 6272 kg/ha of corn was a fantastic yield and so can some of you, if you think back. The average corn yield in Michigan in 1982 was 6837 kg/ha (MDA, 1983). But we still have too many people thinking and saying that no-till won't work. And presently, there are some situations where it will not work the way we want it to.

No-till is the ultimate in conservation tillage. We can and should talk these next two days about other forms of conservation tillage, but I don't think we should settle for less than the best in the long run. That's almost an Un-American attitude. Let's make our long-range objective no-till farming of all crops.

I pondered over this topic for a long time. Conservation tillage, in particular no-till, can be and often is an emotion-laden topic. Most of us are comfortable with the moldboard plow and, admittedly, it has served us very well and will continue to do so in certain situations in the foreseeable future. On the other hand, when and where we can reduce tillage, we should. After all, our objective is to produce an adequate supply of food and fiber at a reasonable cost with a reasonable return (including profit) to the farmer, while we protect the resource base for future generations. Agriculture's objective is not to go out and disturb our topsoil, not to destroy its vegetative cover, and not to subject it to the ravages of wind and water unnecessarily.

Even though the Payment-In-Kind program reduced Michigan's corn acreage about 35 percent last year, we still raised over 40,500 ha of no-till corn. We also raised over 14,200 ha of no-till soybeans, wheat, hay, barley, oats, sorghum, edible beans, sweet corn, asparagus, and green beans. Other types of conservation tillage were practiced on over 445,000 ha in 1983 (USDA, 1983). To put these numbers in perspective, Michigan's cropland currently totals about 3.6 million ha (NRI, 1982).

Why should we be excited about and interested in conservation tillage? For a variety of reasons, I believe, some of which have implications for soil erosion control, energy consumption, wildlife food and cover, water quality, and cost of production. You can probably list some other reasons; I have not tried to be comprehensive but rather provide some food for thought. Obviously, from the viewpoint of the Soil Conservation Service, we are excited because of the implications for erosion, sediment, and agricultural non-point pollution control.

Conservation tillage, and in particular no-till, seems finally to have the potential to bring sheet and rill erosion reasonably under control. Sheet erosion is insidious and difficult to detect with casual observation. Even we in the Soil Conservation Service sometimes talk about sheet erosion but show examples of gully erosion because gullies are easy to see and tillage operations cannot hide them.

Preliminary results from the 1982 National Resource Inventory indicate that about 1.0 million ha of Michigan cropland are experiencing sheet and rill erosion at rates that will not sustain productivity indefinitely. Soil erosion in Michigan is more of a problem now than it was 10 to 15 years ago. Larger fields, more row crops, and farms with little or no livestock are major contributing factors. To be accepted by both farmers and the general public, conservation practices must be practical (that is, fit into the farmers' enterprises), be cost effective, and be environmentally sound. Conservation tillage appears to be our first chance in 50 years to have a long run solution to the sheet erosion problem; but it is not a panacea for all erosion problems, it is another tool, a powerful tool. I'm convinced that as long as we have agriculture, we will need soil and water conservation components in our farm management that evolve along with other changing factors, such as crops, machinery, livestock, and prices.

Three comparisons of soil loss with conventional and no-till practices illustrate what I'm talking about and why we're so excited (Table 1). Believe it or not, at this point I've merely given a preamble to my assigned topic of leadership.

Leadership, like conservation tillage, can also be an emotion-laden topic. Leadership can be a disguise for guarding of turf or prerogatives. But fortunately, for those of us in Michigan, it does not mean that. We have forged a partnership of leaders that needs only nurturing, expansion, and exercise. Our organizing and holding this conference is a result of our partnership. We recognize that the implications and complexities associated with conservation tillage create a job too big and important for any one group, organization, or agency to handle alone or at one organizational level. Many of those who are now providing leadership and need to provide leadership over the long haul are involved in this conference. The partnership of leaders must be exercised continually at the interstate, state, regional, and county levels. We recognize, however, that the real success will be measured at the farm level.

We are dealing with a revolution - a tillage revolution. In 1972, 12.0 million ha were conservation tilled, of

Table 1. Comparison of soil loss with conventional (zero residue) and no-till (3.4 metric tons of residue/ha) for cultivation of a continuous row crop.

Soil Type	Slope %	Length (m)	Soil Loss Conventional (metric tons/ha/yr)	No-till
Blount	4	76	21.0	3.8
Hillsdale	6	76	21.8	4.0
Pipestone	-	--	47.2	0

which 1.3 million ha were no-tilled. Just 10 years later in 1982, U.S. farmers practiced conservation tillage on 40.5 million ha, of which 4.3 million ha were no-tilled (King, 1983).

In preparing for this presentation, I reviewed the publication "Agriculture/2000," a collection of six policy speeches given by then Secretary of Agriculture, Orville Freeman, in 1967. One given to the Annual Meeting of the National Association of Conservation Districts, dealing with "Resources in Action in the 21st Century," discussed soil erosion but did not even mention conservation tillage.

Now, resource conservation is a high priority with Secretary Block. He reiterated his commitment on December 8th, and it is quite common for him to discuss soil erosion and conservation tillage.

The tillage revolution, of course, has not been overnight. In the 1940s, the introduction of plant growth regulators made no-tillage a possibility. Some success with no-till techniques was reported in the 1950s with the development of herbicides to control sod, grasses, and weeds. Fluted coulters and other planter improvements boosted the no-till effort in the 1960s. Significant improvements in chemicals and planters, of course, took place in the 1970s.

So, the stage is set for us to exercise our partnership of leaders. Who does this partnership involve? In preparing for this conference, I made a list. This is always dangerous because someone is sure to be missed, so let's call it a preliminary list. Even a preliminary list is ominous because of its length. But, this is reality:

(1) Michigan Department of Natural Resources
    Divisions of Surface Water Quality, Groundwater Quality, Land Resource Programs, and Wildlife and Fisheries

(2) Michigan Department of Agriculture
    Drains Division, Environmental Resources Division, and Soil and Water Conservation Division

(3) State Soil Conservation Committee

(4) Michigan United Conservation Clubs

(5)   Michigan Association of Conservation Districts
       (Along with individual soil and water conservation
       districts)

(6)   Michigan State University
       Agriculture Experiment Station
       Cooperative Extension Service
       Departments of Agricultural Economics, Agricul-
       tural Engineering, Crop and Soil Sciences, Fisheries
       and Wildlife, Horticulture, and Resource Development

(7)   U.S. Department of Agriculture
       Agricultural Research Service, Agricultural Stabilization
       and Conservation Service, Economic Research Service,
       and Soil Conservation Service

(8)   U.S. Environmental Protection Agency

(9)   Industry
       Chemical, seed, fertilizer, and equipment companies

What are some challenges for this leadership? With
the above list of parties, an obvious major challenge
is to maintain effective communication and coordination
at all levels.

For a variety of crops and soils the most important
challenges are: (1) Improving planters; (2) Improving
plants and seeds; (3) Improving fertilizing techniques--perhaps
the fertilizers themselves; (4) Improving herbicides and
insecticides so they are cheap, effective, and environmentally
sound; (5) Clarifying costs and returns; (6) Determining
environmental impacts of leaving residue, fertilization
techniques, and herbicides and pesticides; (7) Determining
soil temperature and moisture condition implications;
(8) Transfering technology to the farmer; and (9) Determining
soil properties that affect tillage.

The ultimate challenge is to make no-till our "conventional
tillage." Better yet, we probably should work toward
dropping the term conventional tillage from our vocabulary.
What does it mean anyway? Fall moldboard plow, spring
disk, harrow, cultivate--possibly. Spring moldboard plow,
harrow, cultivate--possibly. Disk, harrow, cultipack--possibly.
The point is, "conventional" is already many different
things, as is conservation tillage.

The problem with "conventional" is that it commonly
means "suitable" or "acceptable." So breaking away from
conventional is fraught with uncertainty because a practice
that is unconventional is automatically in the category

of unacceptable or unsuitable. We need to put no-till in the conventional category, be positive in our approach, and concentrate our efforts on improving the practice for a wide variety of crops and soils.

We in American agriculture have had tremendous success in plant breeding, animal husbandry, machinery development, and many other fields of agriculture. With our experience, if we decide to do it, I'm sure that we can have similar success as we cooperatively guide the evolution of conventional till to no-till.

So, in conclusion, the Soil Conservation Service looks forward to continuing work with all of you, our leadership partners, in further developing and carrying out this tillage revolution.

LITERATURE CITED

King, A.D. 1983. Progress in no-till. J. Soil Water Conserv. 38(3):160-161.

MDA. 1983. Michigan Agricultural Statistics. Michigan Department of Agriculture, Lansing, MI 48909.

NRI. 1982. National Resources Inventory. Preliminary Data. Resource Inventory Division, USDA, Soil Conservation Service, P.O. Box 2890, Washington, DC 20013.

USDA. 1983. Conservation tillage survey, Michigan. U.S. Department of Agriculture, Soil Conservation Service, East Lansing, MI 48823.

CHAPTER 3

CONSERVATION TILLAGE SYSTEMS AND EQUIPMENT REQUIREMENTS

William A. Hayes
Fleischer Manufacturing, Inc.
Columbus, Nebraska 68601

INTRODUCTION

Tillage began when man was evicted from the Garden of Eden. According to Genesis 3:23 of the Bible, "Therefore the Lord God sent him forth from the Garden of Eden, to till the ground from whence he was taken," which was about 7,000 years ago. History tells us that the first tillage tool was a stick. A stick was used to open the soil for seed placement and to dig up competing vegetation.

The first plow was a forked stick and was pulled through the soil by the wife and steered by the husband. The role of woman has changed today but the role of the plow is not much different; only the depth it reaches and the degree of vegetation destruction has changed. Early Egyptian drawings of the plow show that the design used thousands of years ago was very similar to the plow used by early American farmers. Until the last 75 years the plow left field surfaces very rough and with only partial burial of competing vegetation and residue from the previous crops. Since tillage began, man has used the plow as the primary tillage tool, but this generation of man is privileged to witness many farmers park the plow and for the first time use other tools for primary tillage, or produce crops with no-tillage at all.

Mechanization of farm operations has been an important part of the "Industrial Revolution." Unfortunately this mechanization of the farm has resulted in excessive: (1) Land clearing; (2) Tillage of the land and destruction of soil cover; (3) Water runoff; (4) Water erosion; (5)

Wind erosion; (6) Traffic and soil compaction; and (7) Contamination of air and water resources.

The most significant of these excesses is loss of our precious topsoil through erosion. The most important soil factor in water erosion that we cannot economically change is the slope of the land. Therefore, percent of slope is a major factor in selecting the right conservation tillage system. According to the U.S. Soil Conservation Service the average annual soil loss from our 167 million ha of cropland is 11.6 tons per ha per year. To many people, that figure does not mean a thing; it's just another statistic. However, if we assign a dollar value to it maybe we can awaken farmers, politicians, and the general public to the devastation that is taking place. According to a 1978 study by Willis and Evans (1977) of Agricultural Research Service (ARS), the fertilizer nutrients in a ton of topsoil have a value of $5.00. That's over seven years old and prices have gone up. The remaining portion of a ton of soil certainly has some value; therefore, I believe I am very conservative in assigning a value of $10.00 per ton for topsoil. On a 200 ha farm, that's 200 x 11.6 x $10.00 = $23,200 per year loss. This is a large sum to the farmer. Now let's see what the loss is nationally - the value of the topsoil lost annually from our 167 million ha of cropland is: 167 million x 11.6 x $10.00 = $19.4 billion dollars. Such losses have been going on for years. In just the past 30 years, the loss would be valued at $582 billion dollars (19.4 billion x 30 = $582 billion). We Americans can't stand a loss like that. It is said that, "as the soil goes so goes the nation." History records a graveyard of great nations resulting from excessive soil erosion and loss of their capacity to produce food. We, too, are headed for the "graveyard" unless we greatly reduce soil erosion. It is difficult to get the attention of people during a period when production is greater than our needs. In the humid part of the U.S. conservation tillage properly applied to the land will eliminate wind erosion and, in combination with other soil conservation practices, can reduce water erosion by 50 to 97 percent when compared with conventional tillage.

Those nations of the past that were conquered by erosion were not troubled with compaction, another villain that plagues the U.S. today. The "other villain" that plagues this country is caused by tillage and equipment traffic. Therefore, to help save our nation, the conservation tillage systems and the conservation tillage equipment selected must result in drastically reducing soil erosion and soil compaction to prevent adding another tombstone to the graveyard of nations. Not only do we have to reduce erosion and compaction but crop yields must be equal or better than with conventional tillage.

System Requirements

Basically there are three different kinds of conservation tillage based on the amount of soil disturbance: (a) No-tillage, zero tillage, or slot planting; (b) Strip tillage which includes till planting and rotary strip tillage; and (c) Non-inversion tillage which includes chisel tillage, disk tillage, and rotary tillage.

No-Tillage. The no-tillage crop production system disturbs about 10 percent of the surface area and leaves a protective quantity of residue uniformly distributed over the soil surface the year around. No-till has four basic operations: fertilize, spray herbicide, plant and harvest. Planting is accomplished with no soil disturbance except in the immediate area of the crop row. Herbicides are applied for weed control, prior to or after planting. Starter fertilizer is usually applied through the planter but the bulk of the fertilizer is knifed into the soil or broadcast on the soil surface. Lime and other amendments are applied on the surface as needed.

Strip Tillage. Strip tillage includes crop production systems where tillage operations for seedbed preparation or removal of the ridge top and planting disturbs about one-third of the distance between the rows, and a protective cover of residue is left on the soil surface the year around. Tillage or skimming off the top of the ridge, planting, and application of herbicides are usually accomplished in one operation. Till planting has five basic operations: chop stalks, fertilize, plant, cultivate and harvest. Herbicides and/or cultivation are used for weed control, and in till planting the ridge is rebuilt annually during cultivation. Till planting, rotary strip tillage, and similar operations are considered to be strip tillage systems. Starter fertilizer is usually applied through the planter but the bulk of the fertilizer is broadcast or knifed into the soil with a separate operation. Lime and other amendments are also applied on the surface as needed.

Non-Inversion Tillage. Non-inversion tillage includes crop production systems which mix crop residues with the top 8 to 10 cm of soil in the entire row and interrow area. A protective quantity of residue is left on the soil surface the year around. Herbicides and/or cultivation are normally used for weed control. Herbicides are normally incorporated with tillage prior to planting. Surface applied and post-emergence herbicides may also be used separately or in combination with the planting operation. The bulk of the fertilizer and lime are broadcast and incorporated into the soil with tillage or knifed into the soil after tillage. Starter fertilizer is normally applied in the planting operation. The chisel system, disk system, stubble mulch system, rotary tillage system, and similar systems are all considered to be non-inversion tillage.

Equipment Requirements

There is an abundance of conservation tillage equipment on the market. Selecting the equipment to accomplish tillage objectives involves some very important evaluation of conservation tillage systems and equipment and planning decisions. The conservation tillage system and equipment should perform the following functions: (1) Modify and improve the micro-environment; (2) Reduce wind and water erosion, and compaction to acceptable levels; (3) Kill competing vegetation; (4) Provide for application of nutrients and other amendments; (5) Insure timely planting; (6) Provide firm seedbed, desired seed placement, seed-soil contact, and cover seed; and (6) Provide soil conditions for timely harvest.

Modify the Micro-Environment. No-till, zero tillage, and slot planting do very little to modify the environment. All no-till planters have a disk in the front to cut the residue. All except Buffalo use a ripple or waffle coulter to cut residue. Buffalo uses a straight coulter which is easier to keep sharp, requires less weight for penetration, and causes less disturbance to the seedbed. Waffle or ripple coulters are necessary on all except the Buffalo Planter for no-till because they have disk openers that require a band of worked soil in order to get penetration of the disk opener. As the planters with disk openers move through the soil, the disk has a tendency to ride the planter out of the ground. Therefore, extra weight may be needed to insure planter penetration under very firm soil conditions. The Buffalo and Cole slot planters have a shoe and a chisel respectively, instead of a disk opener, that have a tendency to pull the planter into the ground. A ripple coulter (2 cm) is better than a waffle coulter (4 cm) for the seedbed because it breaks up less soil and does a better job of cutting residue. If the soil is a little on the wet side, a waffle coulter will throw considerable soil out of the seedbed and sometimes leave little loose soil to cover the seed.

The no-till seedbed is coldest of all conservation tillage systems because residue covers most of the area including the seed zone. This can delay planting, germination, and seedling growth, especially on soils that are naturally slow to drain internally. Burrows and Larson (1962) reported that each ton of residue lowers the soil temperature $0.7^{\circ}F$, and often there are 2 or more tons of residue on the soil surface with no-till.

With till planting or ridge planting the micro-environment is changed by creating and maintaining a ridge at the last cultivation. A special heavy residue cultivator creates a 15 to 20 cm ridge in 76 cm rows and a 25 to 36 cm ridge in 90 to 95 cm rows. The soil on the top of the ridge warms up and dries out quicker than under any other tillage system which permits earlier planting, hastening germination and early seedling growth. Burrows (1963) reports ridges running east-west were $2.5^{\circ}F$ warmer than a conventionally tilled, flat surface. North-south ridges were slightly lower in temperature than east-west ridges but still warmer than conventional. Crop residue left on the surface reduces evaporation. The more residue left on the soil surface, the greater the reduction in evaporation.

Remnants of prehistoric ridged agricultural fields have been uncovered by scientists in Michigan and Wisconsin. It has been determined that they were constructed between the years 1100 and 1300 A.D. by the Indians. The fields range in size and shape up to 40 ha in size. The ridges varied in height from 15 to 76 cm with row widths of 1 to 2 m. Riley and Freimeuth (1979) reproduced such "ridged garden plots" in Richland Community College, Decatur, Illinois. They reported, "Our experiments support our specific hypothesis that their ridge-and-furrow configurations in Wisconsin and Michigan functioned as frost-drainage mechanisms in an area where radiation frosts are a severe limiting factor to length of growing season and therefore to crop production."

Because of the elevation of the top of the ridge, water in the pore space drains to a lower level, giving the desired aerobic conditions in the seed zone. This allows 7 to 10 days earlier planting on the average. Dr. Don Eckert of Ohio State University stated at a recent conservation tillage demonstration that his conventional plots on Hoytville clay were not ready to plant in the spring of 1983 until June 1st. He said the till plant plots could have been planted on three occasions in May, the earliest being May 1st. This means a 30 day head start.

Planter equipment used on the ridge should have a stabilizing coulter on each row that not only holds the planter on the ridge but cuts the residue. As in no-till, the conventional planters can be modified by adding an attachment to the front of the planter to stabilize the planter, cut residue and clear the top of the ridge. Buffalo also makes a pre-plant row conditioner that clears the ridge in a separate operation ahead of planting. This permits the farmer to use any conventional planter with only minor modification. Attachments are made by a number of companies.

Ridges can be no-till planted but the ridges will need to be remade with a cultivator every 2 or 3 years.

Other strip tillage equipment can all be grouped into a type that works or tills the soil on the top of the ridge. One example is the rotary tiller which only utilizes the knives necessary to till a band 25 to 35 cm wide--where the crop row will be planted. I visited with a farmer recently who works the top of the ridge with three waffle coulters; one coulter precedes the disk openers for the seed, liquid nitrogen is knifed in following the second coulter while dry fertilizer is knifed in following the third coulter. Strip tillage planted in a flat surface is also used. The tiller knives are set to work the soil about 8 cm deep.

Non-inversion systems all modify the environment in about the same way. The same principle applies to all; some residue is left on the soil surface and the rest is worked 8 to 10 cm into the soil. They differ in that the disk and rotary tiller are more damaging to soil structure than the sweeps and chisels. The purpose of tilling with these tools is to dry out the soil for planting, bury some residue, incorporate herbicides and amendments, and kill existing vegetation. These tillage operations break the capillary tubes that transport water to the seed zone. If water is not received from above, such seedbeds can dry out and contribute to a poor crop stand.

Ahead of each chisel there must be a coulter to cut the residue. Often another tool called a finisher that breaks up clods and contributes to a finer seedbed trails behind the chisel shanks.

Larger disks are often employed in the disk tillage system. Such disks bury large amounts of residue and are damaging to soil structure. Disks bury more residue

Table 1.   Amounts of residue remaining with different tillage implements (Hayes and Young, 1982).

Implement	Percent Surface Residue Remaining After Tillage
Moldboard plow (18 cm deep)	0-5
Chisel plow 5 cm wide	75
Heavy tandem disk	60
Offset disk	50
Field cultivator (30 to 40 cm sweep)	80

than chisels (Table 1).   Twice over with a disk leaves only 25 percent of the residue on the surface, one of the major reasons why the system is not as effective in erosion control as no-till and till plant.   In addition to leaving less residue on the surface, the disk detaches more soil particles which are readily picked up by water flow.

Reduce Erosion and Compaction.   All conservation tillage systems reduce both wind and water erosion.   The quantity of erosion that occurs is largely dependent on the quantity of residue that is left on the soil surface and the amount of soil particles that is detached in the tillage and planting operation.

The amount of residue left on the soil surface varies with different equipment.   A plow leaves between 0 and 5 percent; a chisel with 5 cm wide points leaves 75 percent, a tandem disk, 50 percent, and a field cultivator, 80 percent (Table 1).   A till planter leaves 70 percent, and a no-till planter leaves about 90 percent on the surface (Table 1).   Note that these percentages are not percent ground cover but the portion of material left on the surface. Yasar and Wittmuss (1976) reported that till plant on a ridge saved more water over winter than any other system, 77.5 percent (Table 2).   Next in order were chisel, chisel disk, chisel plow, no-till, sweep plow, and last, the plow.   Disk and chisel systems normally reduce water erosion about 50 percent; till-plant, 75 percent; and no-till, about 90 percent (Table 3).   Even though no-till is fifth in water saved (Yasar and Wittmuss, 1976), it is first in soil saved.   Till-plant is first in water saved and second in soil saved.

Table 2. Total amount of water conserved from precipitation over the winter (Yasar and Wittmuss, 1976).

Tillage System	Centimeters Saved	Percent Saved
Plow	107	46.3
Chisel Plow	160	69.2
Sweep Plow	132	56.8
Chisel Disk	163	70.9
Chisel Coulter	170	74.0
Till-Plant	178	77.5
No-Till	135	58.2
Precipitation	231	--

All conservation tillage systems practically eliminate wind erosion. Residue on the soil surface reduces wind velocity at the soil surface, thereby reducing the ability of the wind to transport soil particles. All conservation tillage systems reduce compaction compared to conventional tillage. Tillage destroys soil aggregates and the down pressure of tillage tools and wheels of heavy equipment compacts or presses individual soil particles so tightly together that the total pore space in the soil is drastically reduced. Eighty percent of the compaction occurs the first time a wheel passes over the surface of a soil.

Table 3. Estimated soil loss for various tillage systems (Hayes and Young, 1982).

System	Soil Loss (T/ha/yr)
Conventional and Plant	20.2
Disk or Chisel and Plant	10.5
Till-Plant	4.5
No-Till or Slot Plant	1.8

Source: Estimates based on Universal Soil Loss Equation for average conditions in Northern Cornbelt (average soil erodibility, 4 percent slope 6 m long, up and down slope, continuous corn, and 6,900 kg/ha yield).

Little additional damage is done if a wheel passes over the same area 10 times.

Robertson and Erickson (1980) reported the total pore space in a Fox Sandy Loam soil was reduced from 59.8 percent to 41.9 percent by cropping. Air pore space was reduced from 26.0 to 10.5 percent. In the cropped area, it took 40.2 minutes for 100 ml of water to percolate through a soil core while it took only 48 seconds for the same amount of water to pass through a core of the same soil from the fence row in the same field.

Raghavau et al., (1978) of MacDonald College of McGill University, Quebec, conducted compaction studies in 1976 and 1977 that show the effect of compaction on corn yield. They performed the experiment in two successive years: 1976, a wet year; and 1977, a dry year. Total rainfall from May to September was 533 mm (compared to a normal of 432 mm) in 1976, while in the same period in 1977 there were only 375 mm of precipitation. The spring was dry in 1977 with no rain for 3 weeks during the seeding period. They concluded that in 1976, a typical wet year, up to 50 percent losses were attributed to vehicular compaction while in a typical dry year, 1977, such losses were 30 percent.

Surface residue not only reduces wind and water erosion but also reduces water loss. As previously stated, Yasar and Wittmuss (1976) reported that till plant saved 77.5 percent of the 231 mm of precipitation while conventional tillage saved 46.5 percent near Lincoln, Nebraska (Table 2). Note that water loss under no-till was exceeded only by the sweep plow and the plow. Yasar and Wittmuss (1976) reported that each 25 mm of water saved increased yield 542 kg/ha.

Residue on the soil surface not only reduces runoff but reduces evaporation. Harold et al., (1963) found that soil moisture was in greater supply through the cropping season with no-till corn and that it resulted in an increase in corn yield of 570 kg/ha, at the Coshocton, Ohio, Experiment Station (Figure 1). Conventional tillage yielded 6550 kg/ha and no-till yielded 7120 kg/ha, with the increase due to increased moisture availability.

All conservation tillage systems reduce evaporation; the higher the percentage of ground cover, the lower in evaporation. Therefore, equipment that buries residue contributes to loss of valuable moisture.

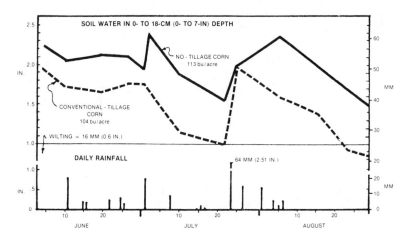

Figure 1. Tillage Systems Soil Water Levels Comparison

Kill Competing Vegetation. Competing vegetation can be killed by herbicides, cultivation, or a combination of the two. Weed control is essential to the success of all conservation tillage systems. Spray equipment, herbicide materials used, and cultivator equipment must be of the highest quality and the management program must be good to insure success.

With strip tillage (till planting or rotary strip tillage), conventional cultivators will not accomplish the task; they will plug with residue. It takes a cultivator with ruggedness and special design to penetrate and move through soil that has not been tilled. This is accomplished by placing a stabilizing coulter on the front of the cultivator or a depth-control tire with a 46 cm coulter behind. Disk hillers are located to the rear and to the side of the coulter or tire. One large sweep per row middle follows in line with the coulter which not only stabilizes the cultivator but cuts the residue so it doesn't catch on the sweep.

An adjustable pitch sweep is attached to a ridged shank, spring loaded for rocky conditions, that maintains uniform and accurate depth control of the sweep. Lack of depth control can result in slabbing of soil and pruning of crop roots. A rigid shank gives better depth control than a spring shank. When a sweep which is mounted on a spring shank meets resistance, the pitch of the shank increases and the sweep goes deeper. Root pruning can be a problem at the second cultivation unless the width of the sweep is reduced from 40 to 25 cm for 71 to 76 cm rows and from 56 to 30 cm for 91 to 97 cm rows.

Conventional cultivators are commonly used where chisel, disk, rotary, and sweep tillage systems are used but heavy residue cultivators do a superior job in cutting residue and in weed control.

Often a 25 to 36 cm band of herbicides is placed over the row and the weeds in the row middles are controlled by cultivation. In the till-plant system this is often the case. However, many farmers also spray the entire area and cultivate only once to maintain the ridge. As previously stated, when farmers no-till or slot plant on a ridge, a special cultivator such as the Buffalo is used at lay-by-time once every 2 or 3 years to maintain the ridge. Such a cultivator effectively controls weeds in no-till when herbicides fail.

Herbicides are usually applied during the planting operation with strip tillage (till-plant and rotary strip tillage). With disk, rotary, and chisel conservation tillage systems, herbicides are normally incorporated with tillage operations. Thorough mixing of incorporated herbicides requires two diskings which buried much residue. From the standpoint of saving surface residues, I believe one disking and use of surface applied herbicides in combination with cultivation will do a superior job of weed control and it will be accomplished more economically. No-tillage herbicide treatments are applied either preceding planting or immediately after planting.

Regardless of the conservation tillage system used, post-emergence herbicides may be necessary on certain occasions. Special shields for protection of crops, spray extensions, rope wicks, and other contraptions are available to help with the job.

Provide for Supply of Nutrients. Nutrients are applied in a variety of ways with each of the systems and by numerous pieces of equipment. Fertilizer nutrients, phosphorus and potassium, and minor elements are normally applied on the surface with all systems; however, deep placement and banding of phosphorus and potassium in both liquid and dry forms is being used and is more nutrient efficient. Machinery for blowing dry fertilizer into a slot has just come on the market in the past few years, although liquid equipment has been available for some time.

Application of dry fertilizer at planting time has not been popular with many farmers because it takes time away from planting. Many of these farmers use a liquid starter even though the cost is higher than dry. They use it because of convenience and reduced labor. Many farmers use no starter, especially if their soil tests are medium to high. However, with no-till and till-plant, starter fertilizer should be used because soil temperatures can be cooler. The insurance of readily available nutrients, when planting early, often pays off.

Nitrogen sources most popular to corn farmers are anhydrous ammonia and 28 percent liquid nitrogen. Research shows that both nitrogen sources should be sealed with soil at the time of application. Commercial applicators are available for rent by farmers, and if equipment or labor is a problem, custom application is an alternative. Many farmers own their own equipment and more and more are purchasing attachments to their heavy residue Buffalo cultivator and side-dressing with anhydrous ammonia during cultivation. Others are removing the cultivator sweeps and installing anhydrous knives in their place--converting the cultivator into an applicator and applying anhydrous ammonia before and after planting. When applying pre-plant ammonia nitrogen most farmers find it best to apply it in the side of the ridge. Others are applying 28 percent liquid nitrogen behind the disk hillers. The liquid nitrogen is then covered with soil moved by the sweep that follows.

Insure Timely Planting. Based on experience in the Corn Belt and the northeast, a wet spring is probably the most common problem interfering with getting spring crops in on time.

Based on soil temperature, agronomy specialists say the average desirable date for conventional corn planting in the Corn Belt is about May 1 and for no-till, about May 5th. Till-plant and rotary strip tillage on a ridge can be planted 7 to 10 days earlier than conventional because the ridge warms up and dries out earlier. Reference is made to an earlier statement of Dr. Erkert of Ohio State University, about their experience at the Hoytville station this year. Therefore, when the season is wet and cold, planting on a ridge offers the best opportunity to get the crop planted at the proper time, year in and year out. The least to the greatest delay in planting will occur in the following order: till plant (no-till) and rotary strip tillage on a ridge, conventional tillage, chisel, disk, and rotary conservation tillage, and no-till (on the flat).

Residue on the surface of the chisel, disk, and rotary conservation tillage systems causes more water to be held by the soil and lowers the soil temperature. Delay with no-till is primarily due to low temperatures caused by surface residue.

Provide Firm Seedbed, Seed Placement, Seed-Soil Contact, and Cover Seed. A firm seedbed is essential to insure a predictable supply of moisture to the seed. The seedbed is located under the seed when the seed is dropped. When soil settles and soil particles that are called capillary tubes. The weight of the atmosphere creates down pressure on the soil and on the water in the soil. This down pressure forces water up through the space between the soil particles, and movement of water up through the tubes is called capillary action. The coarser the capillary tubes, the less distance water will move up the tubes. Therefore, if the area below the seed is tilled and soil particles are too far apart, no moisture moves up from below and the soil around the seed dries out and moisture essential to germination will not be supplied in the seedbed area until rain falls. A firm, moist seedbed is therefore necessary to assure rapid, uniform germination of seed.

Most seedbeds today are tilled below the depth of seed placement. This is definitely the case with conventional tillage, rotary tillage, rotary strip tillage, chisel tillage, and disk tillage. It is often true with no-tillage if the fluted or ripple coulter is set to till too deep. When slot planting or till planting with the Buffalo planter, the straight coulter in front of the opener will not till the soil; there is a minimum of disturbance and a forming bar on the rear of the shoe firms the soil where the seed is placed. Following seed placement is a wheel that presses the seed into the firm soil. This assures moisture movement through the capillary tubes to the seed from three sides. Immediately following the press wheel are two adjustable covering disks that move as much or as little loose soil as desired over the seed - no down pressure is exerted on the soil covering the seed.

Most other planters have disk openers that exert down pressure below and to the side of the seed. If the seedbed moisture conditions are right, there is good seed-soil contact after the covering wheels squeeze the soil around the seed, including on top of the seed. If it is a little too moist, excessive compaction occurs over the seed when this is done. If it's too dry, poor seed-soil contact occurs and the seed germination is delayed until moisture comes from above.

Stout et al., (1961) conducted an experiment at Michigan State University on the "Effect of Soil Compaction on Seedling Emergence Under Simulated Field Conditions." They concluded: "Evidence obtained in the laboratory indicates that planters for sugar beets, corn, and beans should be designed to (a) pack the soil below the seed level, (b) press the seed into the compacted soil, and (c) cover the seed with loose soil. The moisture content of the soil greatly influences the effect of soil compaction on seedling emergence. Packing the soil at seed level may improve seedling emergence if adequate moisture is available below the seed. If moisture is not available below the seed, packing at any level is of no benefit." They also found that when the field surface was too moist, as little as 0.035 kg/cm2 down pressure on top of the soil over the seed interfered with seedling emergence. When moisture was adequate 0.035 kg/cm2 gave the most rapid emergence but 0.352 to 0.703 kg/cm2 surface pressure suppressed emergence. McKibben (1982) found that pressure on a wet soil reduced the stand of corn and cut yield by 2200 kg/ha.

This research is highly significant. Ideally, the conservation tillage system and the equipment selected to produce the crop should be geared to provide soil conditions that comply with the conclusions of this experiment.

Provide Desired Seed Placement. Little space will be devoted to this subject but that doesn't lessen its importance. Most of the equipment on the market, if properly set, will uniformly place the seed at the desired spacing and at the proper depth when soil conditions are ideal. Under adverse conditions, such as hard soil surface or large amounts of residue, those planters with fluted or waffle coulters and disk openers will have difficulty with penetration and depth of planting. Residue is often poked into a soft seedbed contributing to poor seed-soil contact. Under wet soil conditions, the chunks of soil thrown out of the seedbed area sometimes create a coverage problem.

The straight coulter with a depth band, such as that on the Buffalo planter, if kept sharp, will cut the residue and will not poke it into the seedbed. Regardless of the soil conditions, with residue clippers on the planter, residue is always cut in two pieces and maybe as many as four. A straight coulter penetrates hard ground more easily and doesn't throw as much soil as a fluted coulter when the soil is wet.

Under hard ground conditions, the sweep on a till planter or shoe on a slot planter has a tendency to pull the planter into the ground. Better depth control of seed under adverse conditions is thereby attained.

Provide Soil Conditions for Timely Harvest. No-tillage and strip tillage (till-plant and rotary strip tillage) provide the most solid footing and permit the earliest harvest under unfavorably wet fall harvest conditions because the soil is more firm and the soil is in better physical condition and better able to handle the water that falls. No-tillage or slot planting and till planting systems provide those firm soil conditions that permit timely harvest under more adverse wet weather than conventional or other conservation tillage systems.

Production Costs

Conservation tillage systems usually result in lower production costs than conventional tillage. Dickey et al., (1981) reported that fuel and labor requirements are significantly reduced (Tables 4 and 5). Note that no-tillage has the lowest fuel and labor requirements and till-plant on a ridge is second lowest.

Table 4. Diesel fuel requirements (l/ha) for various tillage systems (Hayes and Young, 1982).

Operation	Mold-board Plow	Chisel Plow	Disk	Rotary till	Rotary* Strip Till	Till Plant	No-Till
Chop Stalks	--	--	--	--	--	5.14	--
Moldboard Plow	21.04	--	--	--	--	--	--
Chisel Plow	--	9.82	--	--	--	--	--
Fertilize, Knife	5.61	5.61	5.61	5.61	5.42	5.61	5.61
Disk (first)	6.92	6.92	6.92	6.92	--	--	--
Disk (second)	6.92	6.92	6.92	--	--	--	--
Plant	4.86	4.86	4.86	13.28	8.14	6.36	5.61
Cultivate	4.02	4.02	4.02	4.02	10.85	4.02	--
Spray (2)	--	--	--	--	--	--	4.30
TOTAL	49.37	38.15	28.33	29.83	24.12	21.13	15.52

*Rotary strip till data from a farmer (cultivation includes hilling).

Table 5. Typical labor requirements (hr/ha) for various
tillage systems* (Hayes and Young, 1982).

Operation	Mold-board Plow	Chisel Plow	Disk	Rotary till	Rotary Strip Till	Till Plant	No-Till
Chop Stalks	--	--	--	--	0.42	0.42	--
Moldboard Plow	0.94	--	--	--	--	--	--
Chisel Plow	--	0.52	--	--	--	--	--
Fertilize, Knife	0.32	0.32	0.32	0.32	0.32	0.32	0.32
Disk (first)	0.40	0.40	0.40	0.40	--	--	--
Disk (second)	0.40	0.40	0.40	--	--	--	--
Plant	0.52	0.52	0.52	0.99	0.87	0.62	0.62
Cultivate	0.45	0.45	0.45	0.45	0.45	0.45	--
Spray (2)	--	--	--	--	--	--	0.54
TOTAL	3.01	2.59	2.08	2.15	2.05	1.80	1.48

*Assuming 100 hp tractor and matching equipment for average
soil conditions.

Mowitz (1983) reported on Purdue conservation tillage
research in an article entitled, "Cut tillage and get
more bushels from your production dollar," in the October,
1983, issue of Successful Farming. He shows that till-
plant resulted in the lowest cost of production, $202
per ha, and no-till was next to the lowest cost at $210
per ha. Had the Purdue researchers banded their herbicide
in the till-plant plots, the spread in cost of production
would have been even greater in favor of till planting
on a ridge. Many farmers report that their herbicide
costs range from $16 to $28 per ha by band placement with
the till-plant system. The plow was the highest cost
at $240 per ha. The Purdue research team of Doster et
al., (1983) came to four basic conclusions that are summarized
by Mowitz (1983) as follows: (1) Net returns with till
planting were highest of all tillage systems on all soils;
(2) On light colored, (highly erodible) sloping, well-drained
soils, till-plant and no-till had highest net returns;
(3) On light colored, nearly level, and somewhat poorly
drained soils, till-plant and fall chisel were most profitable
with continuous corn. In rotation programs, spring disking
was second to till-plant in profitability, and spring
plow and no-till were least profitable; (4) On dark colored,

Table 6. Production costs (dollars per hectare) by tillage system (Hayes and Young, 1982).

	Herbicide	Fuel and Labor	Machinery	Labor	Total Cost
Till-Plant	51.87	69.16	76.57	2.47	202.54
No-Till	66.69	69.16	69.16	4.94	209.95
Disk	39.52	71.63	91.39	14.82	217.36
Chisel	32.11	81.51	93.86	12.35	219.83
Plow	32.11	83.98	108.68	14.82	239.59

level, poorly-drained soils, till-plant and fall plow were most profitable for continuous corn. In rotation program, fall chisel was second to till-plant. Spring plow and no-till had the least net returns.

SUMMARY

Based on the information presented, I believe the land in the area with an R value ranging between 1 and 175 (Figure 2), and with slope of between 0 and 7 percent average slope length is best suited for till plant on ridges (contour and terraces for long slopes); between 7 and 12 percent slopes, till plant on ridges on contour and if contouring is not possible, use no-till up and down the slope; about 12 percent and less than 18 percent slope use no-till; above 12 percent slopes on the contour, chisel tillage, disk tillage, or rotary tillage can also be used with high residue crops (or cover crops) and other needed soil conservation practices. Where R values are between 175 and 350, all information remains the same except the range in slope 0 to 7 becomes 0 and 5 percent slopes. For areas where R values are between 350 and 550, reference is made to Table 6. Where hay and small grain are a part of the crop rotation and only one year of row crop is grown, use chisel, disk, or rotary tillage and other conservation tillage practices in place of till-plant or grow continuous crops in one area on ridges while growing small grain and hay on the rest of the farm.

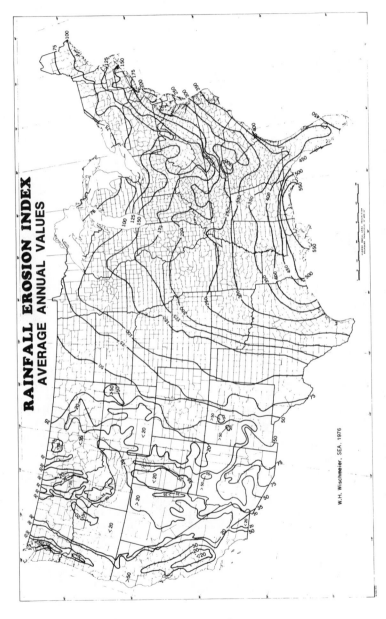

Figure 2. Average annual values of rainfall in the United States.

LITERATURE CITED

Burrows, W.C.   1963.   Characterization of soil temperature distribution from various tillage-induced micro-reliefs. Soil Sci. Soc. Am. Proc. 27(5):350-353.

Burrows, W.C., and W.E. Larson.   1962.   Effect of amount of mulch on soil temperature and early growth of corn. Agron. J. 54:19-23.

Dickey, E.C., A.L. Rider, and P.W. Harlow. 1981. Tillage systems for row crop production. Nebguide G80-535. Agri. Ext. Ser. Univ. of Nebraska, Lincoln, NE.

Doster, D.H., D.R. Griffith, J.V. Mannering, and S.D. Parsons. 1983.   Economic returns from alternative corn and soybean tillage systems in Indiana. J. Soil and Water Cons. 38:504-508.

Harold, L.L., G.B. Triplett, Jr., and W.M. Edwards.   1963. No-tillage Corn: Characteristics of the System. Ohio Agri. Exp. Sta. Mimeo.

Hayes, W.A., and H.M. Young, Jr. 1982. Minimum Tillage Farming/No-Tillage Farming. No-Till Farmer, Brookfield, WI. 197 pp.

McKibben, G.E.   1982.   Herbicides for 0-till corn in sod, 1981.   Update 82, a research report of the Dixon Springs Agricultural Center. pp. 17-19.

Mowitz, D.   1983.   Cut tillage and get...More bushels from your production dollar. Successful Farming (October):10D.

Raghavau, G.V.S., E. McKeyes, F. Taylor, P. Richard, and A. Watson.   1978.   Corn production loss in successive years by wheel traffic compaction. ASAE paper 78-1533.

Riley, T.J., and F. Freimeuth.   1979.   Field systems and frost drainage in the prehistoric agriculture of the Upper Great Lakes. American Antiquity 44(2):271-285.

Robertson, L.S., and A.E. Erickson. 1980. Compact soil-visual symptoms.   Michigan State University Cooperative Extension Bulletin E-1460, November.

Stout, B.A., W.F. Buchele, and F.W. Snyder. 1961. Effect of soil compaction on seedling emergence under simulated field conditions. Agri. Engineering (February):68-87.

Willis, W.O., and C.E. Evans. 1977. Our soil is valuable. J. Soil and Water Cons. 32:259-260.

Wischmeier, W.H., and D.D. Smith. 1978. Predicting rainfall erosion losses: a guide to conservation planning. USDA, SEA Agri Handbook No. 537. pp. 6 and 22.

Yasar, A. and H.D. Wittmuss. 1976. Moisture use and crop yield under various tillage treatments. Mimeo, Agri. Eng., Univ. of Nebraska, Lincoln, NE.

CHAPTER 4

RIDGE TILLAGE FOR EROSION CONTROL

Ernest E. Behn
Boone, Iowa   50036

There is hardly anything in the world so powerful as an idea whose time has come. This "idea" centers around the process of conserving our soil resources by leaving all crop residues on top of the ground year after year after year. We believe if we take care of our soil today we will have food for tomorrow. In the past this "idea of leaving residues" has produced a negative reaction in the minds of most of the people who were exposed to it, but this use of residues is the most economical way to do it. Not just farmers, but Extension Service people, researchers, Soil Conservation Service employees, and others reacted in a negative way. Not all, of course, but most of these people questioned the wisdom of leaving residues on the surface. And I guess you might even say this was a reasonable reaction because farmer experience and early research demonstrated many problems with this reckless method!

About 30 years ago, the late Dr. George Scarseth at the University of Indiana and Dr. George Browning at Iowa State University independently attempted to raise corn in the previous year's corn residue without plowing. They were not very successful because the necessary herbicides and machinery did not exist. About this time I started working for the Soil Conservation Service, and I was sent to Coshocton, Ohio, for a quick review of Soil Conservation. They had some hills with 10 percent slopes that lost 67 metric tons of soil per ha year after year. By contouring, losses were reduced 50 percent to 33.6 metric tons per ha annual soil loss. By terracing, losses were reduced 75 percent to 15.7 metric tons ha/yr. By leaving all residues on the surface, losses were reduced to less than

one ton per acre per year. So I said, "Wow! This is
great! All I have to do is tell farmers to leave all
residues and my work will be done!" But then I was told,
"You can't do that! Residues leave the ground cold and
wet, planting is delayed, weed control is tough, you get
a poor stand, and bugs will eat your corn!!" This is
the first time I ever heard this, I believed it, I had
no reason to doubt it. We were all programmed to fail
because all the news on the subject was bad. We were
told over and over again the cold-wet-ground, bugs-will-
eat-your-corn, theory. This came at us like a broken
record, repeated over and over again, and it was pretty
well set in our minds.

Well, after all that, no one in his right mind would
want to give this method a try, and if they did it was
a feeble effort which they expected to fail, and if you
expect it to fail it probably will. As I just mentioned,
all through this early period, 20 to 40 years ago, everyone
recognized the great erosion control benefit of residues.
Researchers found that residues, combined with contouring
and terracing could give almost 100 percent control of
wind and water erosion instead of 50 or 75 percent that
was experienced with contouring or terracing alone. This
was no secret. We just knew that we couldn't use the
residues for this purpose because it would give us all
those troubles, and it just wasn't a suitable alternative.
I say we, because I, too, was convinced it wouldn't work.
I, too, listened and was affected by this bad news, repeated
over and over again. We were problem oriented, and this
is where I was when I started working for the Soil Conservation
Service, and this is how I felt when I bought a small
farm to live on and operate a few years later.

Like a good farmer, I fall plowed my little farm,
I disked and harrowed to prepare a seed bed and like a
good conservationist I contoured, and built terraces and
waterways. However, as time went on I began to see that
this "conventional" farming had problems, too. I only
had about 50 ha but could hardly get all the work done
on weekends and evenings; when the wind blew, dust was
carried into the air; and when it rained, silt filled
my terraces and when it didn't rain, the soil dried out
and cracked wide open. Corn borers ate my corn, blight
reduced my yield, and weeds were hard to control. All
this is the absence of residues and in spite of the fact
I had carefully plowed all the residues under.

At this time, on a visit to Arizona, I saw farmers
planting cotton on top of carefully prepared ridges.
These ridges were eight to ten inches high and were designed

to permit water to flow slowly in the valley between the rows (ridges) and thus irrigate the crop. I began to visualize planting corn and beans on the top of ridges like this up in Iowa. In the fall, cornstalks and soybean residues would fall down in the valleys and the tops of the ridges would protrude up out of the residues and warm up and dry out quickly in the spring. I began to feel that the freezing and thawing and weathering of this ridge during winter would produce a nice, mellow but firm seedbed in which to plant corn or beans in the spring. It would always be firm, moist, and free of clods. The more I thought about it the more I felt ridges would solve at least some of the problems of planting in residues left on the surface of the ground.

Now for the good news! For the past 17 years we have planted on such ridges, and the results have been fantastic and greatly exceeded our expectations. Now I can say without reservation....this is an idea whose time has come. There is simply no reason to hesitate....for the ridges do solve 95 percent of the problems normally associated with planting "no till." The soil does warm up quickly in the spring, and tops of the ridges dry out quickly to permit early planting. Residue in the valleys prevent both wind and water erosion and prevent water from running to low spots that are normally too wet. There is a dramatic saving in time, labor, and machinery costs, and fuel savings up to 50 percent. In many cases, ridges on the contour make costly terraces unnecessary. A terrace every 76 cm has got to be better than a terrace every 61 or 91 m. If the hillsides are very steep and irregular there is a possibility that terraces may be needed also, but when they are used, they won't fill up with silt and are much easier to maintain. This is all good news, and our yields have been far above the average for the community. In the last three years our yields in the Boone County Corn Yield Contest have been 11,164, 12,606, and 11,603 kg/ha. The average for the county has been 6900 to 7840 kg/ha in the same period.

Truly the time has come for us to forget our prejudices and go full speed ahead. Most farmers, and researchers, too, do not really realize what it is like not to have to get ground ready in the spring....and they don't realize that when you are through harvesting in the fall you are through! There is no more fall work necessary. I soon began to realize that the "best seedbed in the world is created by doing nothing." The ridges are made at cultivating time in midsummer, then just leave them be. Don't run on them with wagons, trucks, combines or manure spreaders. Just don't touch them, leave them be until spring and the best seedbed you can imagine will be right there waiting for the planter. It will be firm, moist, and mellow. The planter will cut off 2.5 cm of the ridge at planting time and leave a smooth, moist row, perhaps 25 cm wide, that is ideal for a band of herbicide. It seems to me herbicides work better on this smooth moist band than they do on loose, dry soils such as exist when soil is disked before planting. From now on planting will never be done in loose, dry soil or a cloddy seedbed. A loose, dry, cloddy seedbed is for farmers who disk or use field cultivators ahead of planting. Herbicides that need to be incorporated were invented for those who "work up" a seedbed before planting.

Now when someone says they would like to start a no-till ridge system but they wonder about the problems, I know they assume that when they farm conventionally they don't have any problems....and this is a big myth. The problems of "conventional" farming are numerous and difficult to solve. When they are listed and exposed it is indeed amazing that anyone continues to farm this way....especially when the no-till ridge system is easily available to all. Here is a partial list of conventional farming problems: (1) it is difficult to control water and wind erosion (bare soil); (2) it takes twice as much time as no-till ridge; (3) it takes too much machinery; (4) fuel costs are high, twice as much as no-till ridge; (5) moisture evaporation from bare soil is high; (6) big equipment causes compaction; (7) corn is a problem in beans (beans after corn); (8) weed control is difficult (150 years of plowing, disking, and harrowing, and we still fight weeds); (9) bugs will eat the corn just like they have in the past (corn borer, chinch bugs, stalk borer, cut worms); (10) bare soil looks bad when you know it is not right.

There are other problems, but the above list should discourage anyone who really thinks about it from using conventional clean till methods.

We put herbicide in the row at planting time and if necessary broadcast additional herbicide just before or right after planting. We broadcast phosphorus and potassium on top of the ground and knife in nitrogen between the rows, but it all could be put on as liquid. Fertilizer can also be put "in the row" at planting time or with the cultivator later. We cultivate once to control weeds and mix residues and fertilizer together. We cultivate a second time to make a ridge for next year. Ridging can be done with disk hillers or with wings on the back sweeps. This is a special cultivator designed to go through residues. A planter equipped to cut off about 2.5 cm of the very top of the ridge is also necessary. There are two no-till planter on the market that do this, but some conventional planters can be converted to do this also. Equipment or attachments are available so there is no longer an excuse for not getting started.

Nothing is more gratifying than to see the explosion that is taking place in the no-till ridge method. People everywhere are now asking questions, and I get phone calls from all over asking, "How is the best way to get started?" Book sales are better than ever, and no-till clubs are sprouting up in many places. Some use the slogan, "SAVE SOIL - TOIL - OIL" for their meetings. Some say this is the "NO-TILL-AGE" and time to make the change. I guess you might say the old way is not always the best way.

So let me repeat, nothing is so powerful as an idea whose time has come; the idea is to make high ridges and then leave all of last year's residues on top of the ground. Success with this method pays off in time, money, and the highest degree of soil conservation. For this idea, its time has come. This is the way to conserve soil today so we can produce food for the future.

CHAPTER 5

DIFFERENCES IN CROP YIELDS AS A FUNCTION OF TILLAGE SYSTEM, CROP MANAGEMENT AND SOIL CHARACTERISTICS

Donald R. Griffith and Jerry V. Mannering
Department of Agronomy
Purdue University
West Lafayette, Indiana 47907

INTRODUCTION

Farmers in Indiana and other Corn Belt states are changing tillage practices more rapidly than they were 10 or even 5 years ago. Current interest in conservation tillage is due to many reasons, but primary on that list is a search for ways to increase net profit. While reduced tillage offers an opportunity to reduce production costs on many farms, yield differences due to tillage practices (if they exist) often have a greater influence on net profit (Doster, 1983).

Evaluating the effect of specific tillage-planting systems on yield is not a simple task, since several associated practices such as pest control, form of fertilizer and method of application, and equipment adaptation, in addition to soil properties, may influence crop response to tillage. However, research and farmer experience in the eastern Corn Belt during the past twenty years have identified the major factors which influence the success of reduced tillage systems (Griffith, 1978). These include soil drainage, previous crop, length of growing season, proper nitrogen application, pest problems and operator management skills.

SOIL DRAINAGE

Most research and experience in the Corn Belt indicates that as soil drainage improves, the need for tillage decreases.

Table 1. Corn yield response to tillage systems, Northern Indiana (Griffith, 1982).

	Well drained Tracy sandy loam	Poorly drained Runnymede loam
	- - - - - - - kg/ha - - - - - - -	
Moldboard plow, disk twice, plant	7,658	8,411
Chisel, field cultivate, plant	7,846	8,160
Till-plant in last year's ridges	8,662	8,348
No-till plant	7,783	7,219

This is illustrated by results from a 7-year continuous corn study in northern Indiana (Table 1; Griffith, 1982). Tillage experiments were conducted on a well drained, sandy loam (Tracy) and on a poorly drained, dark loam (Runnymede).

While all no-plow systems were as good or better than moldboard plow tillage on the well drained soil, no-till yields were reduced on the poorly drained soil. Lower soil temperatures and excess wetness early in the growing season are common on poorly drained soils, and both problems are accentuated when the prior year's crop residues are left on the surface and the soil is not loosened with tillage.

PREVIOUS CROP

Several long term studies have shown the positive effect of crop rotation on corn yields, and the same response is often found with soybeans (Mannering, 1981). The positive influence of rotation is even more important when no-till planting on poorly drained soils, as indicated in results from a 9 year study near Lafayette, Indiana (Tables 2 and 3).

No-till corn yields averaged 942 kg/ha less than plowed yields for nine years of continuous corn, but only 314 kg/ha less when corn followed soybeans. No-till soybean yields were reduced by 404 kg/ha for continuous soybeans,

Table 2. Corn yield response to tillage and previous crop, Chalmers silty clay loam, Lafayette, Indiana (Unpublished data).

Tillage System	Continuous Corn		After Soybeans	
	1980–83 average	1975–83 average	1980–83 average	1975–83 average
	--------- kg/ha ---------			
Fall plow	10,545	10,357	11,299	10,734
Fall chisel	9,980	9,918	11,110	10,608
Till-plant	10,357	---	11,487	---
No-till	8,976	9,415	11,048	10,420

but only 201 kg/ha when soybeans followed corn. Mono-culture reduced yields with all tillage systems, but the reduction was usually greater for no-till planting than for systems including some tillage. Other experiments have shown a similar no-till response to crop rotation when corn follows sod. Note that till-planting compared favorably with plowing in both rotation and monoculture.

Several factors may be involved in the results of these studies. Soil physical properties near the surface

Table 3. Soybean yield response to tillage and previous crop, Chalmers silty clay loam, Lafayette, Indiana (Unpublished data).

Tillage System	Continuous Soybeans		After Corn	
	1980–83 average	1975–83 average	1980–83 average	1975–83 average
	--------- kg/ha ---------			
Fall plow	3,632	3,430	3,833	3,564
Fall chisel	3,363	3,228	3,632	3,564
Till-plant	3,430	---	3,699	---
No-till	3,295	3,026	3,497	3,363

often improve with shallow or no-till planting when corn
is rotated with other crops (Griffith, 1977). Reduced
residue cover after soybeans, and the moisture transpired
by letting sod grow until corn planting, both lead to
improved soil drying and warming. Rotating crops may
also provide fewer pest problems by interrupting the life
cycle of pests which are not controlled by pesticides.

Since residues are concentrated closer to the seed
with no-till planting or shallow tillage, the possible
toxic effect of this decaying residue on germination and
seedling growth of the next crop (allelopathy) is currently
receiving much attention. This effect has been documented
in greenhouse studies when corn follows corn, but its
importance in the field is not clear. An allelopathic
effect when no-till corn follows corn could contribute
to reduced plant growth and yield.

LENGTH OF GROWING SEASON

Tillage systems which leave most of the soil surface
covered by residue have generally been successful in the
central and southern Corn Belt, and the states further
south, on well drained soils. In Wisconsin, Minnesota,
and Canada, trials have often shown reduced yields when
surface residue systems are compared to clean-tilled systems
(Amemiya, 1977; Ketcheson, 1977). In Indiana, results
are usually better in the southern than the northern part
of the state when planting in heavy residues. Reduced
soil temperatures, caused by the residue cover, apparently
have greater effect on plant growth and yield in the more
northern latitudes. Delayed planting, which allows soils
to become drier and warmer, would be more likely to reduce
yield potential in northern latitudes than in southern
latitudes because of the shorter growing season.

PROPER NITROGEN FERTILIZATION

Surface application of nitrogen fertilizer (usually
28 percent liquid) is common with no-till systems and
is often used with till-planting. This nitrogen is subject
to loss through volatilization from urea, through denitrification of nitrates, and through nitrogen use by bacteria
in decaying crop residues (Mengel, 1982). All of these
potential loss situations are aggravated by no-till planting
or till-planting.

The yield and ear leaf analysis data reported in
Table 4 (an average of seven experiments) show that nitrogen
was used most efficiently when injected beneath the soil

Table 4.   Effect of N source and placement on no-till corn yield and ear leaf N, Indian[a] (Mengel, 1982).*

N treatment	Yield (kg/ha)	Ear leaf N (%)
NH_3 injected	8,710	3.06
28% liquid injected	8,490	2.85
28% liquid surface	7,400	2.48
Urea surface	7,720	2.57

*Average of 7 experiments conducted from 1978 through 1980.

surface under a no-till environment.   Applicators are now commercially available to do this.   Where nitrogen is surface applied and broadcast at planting time on no-till corn, the rate of application may need to be increased by 15 to 20 percent in anticipation of extra losses.

PEST PROBLEMS

Weed, insect, disease, and rodent control sometimes becomes more difficult or more expensive or both as tillage is reduced.   During the past 15 years, many yield reductions reported by farmers were caused by poor pest control, especially weed problems.   Technology available now allows control of most pests with conservation tillage.   However, this technology is often more expensive than control methods which include more tillage, and it requires greater skills in recognizing pest problems and applying proper chemicals.

SKILLS AND ATTITUDE

Attitude of the landlord, farmer, and hired labor towards a new tillage system is also important.   All must realize the advantages to be gained with any new system, be willing to learn the new skills needed, and take time to "fine-tune" the system to their specific set of equipment and soils, if maximum yield potential of the reduced tillage system is to be fully realized.

MATCHING TILLAGE TO SOIL

Most tillage research in the Corn Belt shows that response to tillage is related to soil characteristics. A logical first step, then, for farmers who plan to change tillage systems would be to choose one that is well adapted to their soils.

To help farmers choose a tillage system, several states have rated the adaptability of certain tillage-planting systems to groups or classes of soils. In Ohio, soil series were placed in five tillage groups according to soil properties and their influence on no-till planted corn (Triplett, 1973).

For Indiana, Purdue University has published a guide for matching tillage systems to soil types (Galloway, 1977). All named soil series are placed in one of 23 soil-management groups based on drainage and texture. Then, eight different tillage-planting systems are rated as to their adaptability to soils in each of the 23 groups. Tillage systems range from fall plow, conventional tillage to no-till planting and are rated from 1 (well adapted) to 5 (not adapted). Both yield potential and erosion potential were considered in making the ratings. Originally, the ratings were for corn after corn, but they have recently been modified for corn after soybeans, also.

Table 5 gives three example soils which are often found on central Indiana farms and the tillage system

Table 5. Adaptability of tillage systems to three Indiana soil series (Galloway, 1977).

Soil Series[a]	% Slope	Tillage System					
		Fall Plow	Spring Plow	Fall Chisel	Spring Disk	Till-Plant	No-till
				Rating[b]			
Brookston	0-1	1	3	1	2	1	3(2)
Crosby	1-3	2	2	1	2	1	2(1)
Miami	0-6	5	2(3)	2(3)	2	1	1

[a]Brookston--poorly drained, Crosby--somewhat poorly drained, Miami--well drained.

[b]1 = well adapted, 5 = not adapted. Ratings for continuous corn except numbers in ( ) for corn after soybeans.

ratings for these soils. Brookston is a poorly drained, dark colored, silty clay loam. Crosby is a nearly level, somewhat poorly drained silt loam with two to three percent organic matter, while Miami is a well drained silt loam with slopes often exceeding six percent. Note that no-till is rated down on the poorly drained soil due to lower yield potential and plowing and chiseling are rated down on the sloping soil due to greater erosion potential.

Tillage system ratings for corn after soybeans are given in parentheses where they differ from corn after corn ratings. In general, no-till ratings improve on poorly drained soil since residue cover is less than after corn, and systems with full width tillage are rated lower on sloping soils due to less protection from erosion. Note that the till-plant system receives good ratings on all three soils.

YIELD COEFFICIENTS

While the ratings described above are quite helpful in choosing which tillage system to consider, they do not provide a means of actually quantifying yield potential. In response to this need an interdepartmental group of research and extension specialists at Purdue University, using previous tillage system ratings as a guide, have suggested yield coefficients for different tillage systems (Doster, 1983). The coefficients are not based on yield comparisons in any one experiment, but they do reflect research and experience throughout the eastern Corn Belt.

Yield coefficients for six tillage-planting systems are given for both continuous and rotational corn (Table 6). Tillage-planting systems include fall plow, fall chisel, spring plow, spring disk, till-plant and no-till. Within each table, coefficients are given separately for three groups of soils -- poorly drained, somewhat poorly drained, and well drained. The soils groups are also defined in more detail in table footnotes.

Within each soil group for any given tillage system, rotational corn has greater yield potential than continuous corn. Also, for a given soil group and cropping sequence, tillage systems differ in yield potential. The conservation tillage systems (chiseling, till-plant and no-till) show greater yield potential than "conventional tillage" (fall plow) on those soils (Group III) where wind and water erosion control are most important.

These yield coefficients reflect average early-May planting dates. As planting is delayed into late May

Table 6. Corn yield coefficients expected for various crop rotations, tillage systems and soil types in Indiana (Doster, 1983).

Cropping Sequence and Tillage System[a]	Expected Yield Coefficient for Corn on Soil Group[b]		
	I	II	III
Continuous Corn			
Fall plow	1.00[c]	1.00[c]	1.00[c]
Fall chisel	0.97	1.02	1.05
Spring plow	0.93	1.00	1.05
Spring disk	0.95	1.00	1.05
Till-plant (ridge)	1.00	1.02	1.10
No-till	0.90	0.95	1.10
Rotation Corn			
Fall plow	1.07	1.07	1.07
Fall chisel	1.07	1.07	1.13
Spring plow	1.00	1.07	1.13
Spring disk or field cultivate	1.07	1.07	1.13
Till-plant (ridge)	1.07	1.07	1.18
No-till	1.05	1.07	1.18

[a]Tillage system descriptions include:
   Fall plow--fall moldboard plowing, 1-3 spring passes to prepare seedbed.
   Fall chisel--same as fall plow, except a chisel plow is substituted for the moldboard plow. An offset or heavy tandem disk system would have similar yield coefficients.
   Spring plow--same as fall plow, except moldboard plowing is done in the spring.
   Spring disk or field cultivate--1-3 spring passes with a disk or field cultivator to prepare seedbed.
   Till-plant--planting into wide tilled strips on pre-formed ridge tops (no other tillage operation at planting).
   No-till--planting into very narrow tilled strips through old-crop residue (no other tillage operation).

[b]Soil group descriptions include:
   I--Dark, poorly drained silty clay loams to clays, 0-2% slope. Examples: Brookston and Chalmers.
   II--Light (low organic matter), somewhat poorly drained silt loams to silty clay loams, nearly level. Examples: Fincastle and Blount.

or later, the spring disk, till-plant and no-till systems compare more favorably with plow systems. However, with earlier planting, these no-plow systems compare less favorably with plow systems.

Less is known, in comparison to corn, of the true relationship among tillage system, soil type and soybean yield due to the limited amount of research and practical experience available at this time. Knowledge available indicates that the same trends illustrated for corn after corn also apply for soybeans after corn (Table 6). Yield potential for continuous soybeans is reduced for all tillage systems. As planting date is delayed, no-till soybean yield potential improves relative to other systems. No-till double-crop soybeans have consistently produced better yields than other tillage systems used for double-cropping, even on poorly drained soils.

To use the table to estimate expected long-term yield for changing tillage systems, let fall plow be the "standard" practice with a coefficient of 1.00 and assume (or determine) an actual yield per acre for this practice in a particular soil-rotation situation. Then, multiply this yield by the coefficient given for the new system.

Table 6 notes continued.

III--Light, well drained, shallow terrace soils, sands, sandy loams and silt loams with 3% or greater slope, i.e., most soils that are subject to wind or water erosion and/or drought. Examples: Bedford and Fox.

Soils not included in these groups are mucks, bottomlands, dark sands with high water table, and light, flat soils over fragipans (like Clermont and Avonberg). See AY-210, "Adaptability of Various Tillage-Planting Systems to Indiana Soils" for information relating tillage to specific soil series for corn production.

[c]Fall-plow tillage system with early-May planting is used as a reference point (1.00) for each soil group, but actual yield potential may be different between soil groups. As planting is delayed, the spring disk and no-till systems compare more favorably with the plow systems in soil group I. With earlier planting (April) in soil group I, these no-plow systems compare less favorably than shown.

For example, if fall plowing on a Group III soil produces 7,440 kg/ha of corn, switching to no-till planting should increase yield to 8,184 kg/ha (1.10 x 7,440 kg/ha)

These yield coefficients are intended for use by farmers, farm managers, economists and tillage system modellers where on-site yield information is not available. To place most of the soils in Indiana into just three groups for tillage system ratings is, of course, a major oversimplification. Although trends shown by the ratings reflect current knowledge, tillage system relationships for individual soils may vary from those shown. Where local data or experience indicate different relationships, they should be used.

SUMMARY

Tillage system yield trials generally reflect the soil and climatic conditions, and the cultural practices under which the trial was conducted. The relation of these factors to reduced tillage success have been fairly well documented across the Corn Belt. Our knowledge of the relationship of soil drainage and crop rotation to tillage system yields has allowed the assignment of ratings and yield coefficients to tillage systems. These ratings and coefficients are intended to allow long term budgeting in the absence of specific, on-site yield information, but are not intended to predict yield in any specific year.

LITERATURE CITED

Amemiya, M. 1977. Conservation tillage in the Western Corn Belt. J. Soil and Water Cons. 32:29-36.

Doster, D.H., D.R. Griffith, J.V. Mannering and S.D. Parsons. 1983. Economic returns from alternative corn and soybean tillage systems in Indiana. J. Soil and Water Cons. 38:504-508.

Doster, D.H., S.D. Parsons, D.R. Griffith, J.V. Mannering, D.B. Mengel, R.L. Nielsen and M.L. Swearingin. 1983. Estimating potential yield for corn, soybeans and wheat. ID-152, Coop. Ext. Ser., Purdue Univ., W. Lafayette, IN.

Galloway, H.M., D.R. Griffith and J.V. Mannering. 1977. Adaptability of various tillage-planting systems to Indiana soils. AY-210, Coop. Ext. Ser., Purdue Univ., W. Lafayette, IN.

Griffith, D.R., J.V. Mannering, D.B. Mengel, S.D. Parsons, T.T. Bauman, D.H. Scott, C.R. Edwards, F.T. Turpin and D.H. Doster. 1982. A guide to no-till planting after corn or soybeans. ID-154, Coop. Ext. Ser., Purdue Univ., W. Lafayette, IN.

Griffith, D.R., J.V. Mannering and W.C. Moldenhauer. 1977. Conservation tillage in the eastern Corn Belt. J. Soil and Water Cons. 32:20-28.

Ketcheson, J. 1977. Conservation tillage in eastern Canada. J. Soil and Water Cons. 32:57-58.

Mannering, J.V. and D.R. Griffith. 1981. Value of crop rotation under various tillage systems. AY-210, Coop. Ext. Ser., Purdue Univ., W. Lafayette, IN.

Mengel, D.B., D.W. Nelson and D.W. Huber. 1982. Placement of nitrogen fertilizers for no-till and conventional till corn. Agron. J. 74:515-518.

Triplett, G.B., Jr., D.M. Van Doren, Jr., and S.W. Bone. 1973. An evaluation of Ohio soils in relation to no-tillage corn production. RB-1068. Ohio Agr. Res. Development Center, Wooster, OH.

CHAPTER 6

AN OVERVIEW OF THE OHIO CONSERVATION TILLAGE RESEARCH

Glover B. Triplett, Jr.
Department of Agronomy
Mississippi State University
Mississippi State, Mississippi 39762

David M. Van Doren, Jr.
Department of Agronomy
Ohio Agricultural Research and Development Center
and
Ohio State University
Wooster, Ohio 44691

INTRODUCTION

While this is not a historical presentation, no paper with this title should ignore those from Ohio who contributed to conservation tillage before what might be deemed the modern era. The modern era began when herbicides were developed that permitted substitution of chemicals for cultivation in weed control. Before this time, Faulkner (1943) developed concepts that he advanced in his book, Plowman's Folly. Lloyd Harrold, at Coshocton, did some innovative work with tillage on the watersheds there. Bob Yoder, Harold Borst, and Henry Mederski worked at various times on development of trash-mulch seeding techniques. During this time, Dr. Ray Cook in Michigan worked on plow-plant techniques and Davidson and Barrons at Dow in Midland did early and successful chemical seedbed preparation. Each of these made a contribution and helped shape the course of future developments.

The modern tillage era began in Ohio in 1960. This was the same year that no-tillage work was initiated in Virginia and one year after Free, Fertig and Bay (1963) began their studies in New York. That first Ohio experiment consisted of corn planted in a recently established timothy-alfalfa sod. The best herbicide treatment was dalapon plus amitrole applied 3 weeks before planting; none of the treatments included a residual herbicide. By rights this experiment should have failed. We talk about Murphy's law, "If anything can go wrong, it will." But this axiom has another face, "During a given year anything can work," that can get research workers in trouble. Probably due to an extreme set of fortunate circumstances, yields from the best unplowed treatment equalled the plowed and we continued to work. Dave Van Doren and I worked together on the tillage projects and were able to interest others in working with us on various parts of the program.

Organizers of this conference are to be commended on their adopting a systems approach to tillage. Crop management involves identification of factors that influence crop productivity and, so far as possible, adjusting these factors to optimum levels. If one factor is limiting, the entire system suffers. While tillage per se is not a primary requirement for crop growth, it can have a profound influence on several of the needs of the crop. A considerable part of the effort in the early part of the Ohio tillage research program was devoted to the development of functioning crop management systems. After the initial experimental success, areas that were thought to be important for expanded research effort were identified. These included: (1) planting equipment and techniques; (2) fertility, rates and placement; (3) control of weeds and other pests; (4) soil characteristics and response to tillage; (5) soil erosion and pesticide movement with different systems; (6) mulch cover effects; (7) crops and varieties, response to tillage; (8) long term effects of tillage systems; (9) costs and returns from tillage systems; (10) pasture renovation with reduced tillage; and (11) multiple cropping systems.

Some of the anticipated problems or responses failed to materialize while other unanticipated matters that limited the functioning of the systems arose and had to be addressed. During the more than 20 years since the work was initiated, commercial planters have become available, management systems developed, and reduced and no-tillage practices adopted by commercial producers. There also appears to be a shift toward less tillage so that few farmers today would be following the conventional systems of Ray Cook's day.

Various topics of the research list above are subjects for other speakers in this conference and most likely their findings are similar to those from the Ohio studies.

This paper will address several topics that may be of common interest and that are not represented in titles of other speakers.

MULTIPLE CROPPING

In multiple cropping systems, more than one crop is harvested during a given season, intensifying land use, when compared to production of a sole crop planted at the optimum time. Yields of the second crop are usually less. Even so, such systems may be profitable because land charges are spread over two crops. Reduced or no-tillage systems are important in multiple cropping. When used as a management tool, they become a means to an end. No-tillage planting of the second crop directly into the stubble of the first conserves soil moisture and facilitates rapid crop establishment. Two multiple cropping systems may be adapted to parts of Michigan.

Soybeans following Wheat

A soybean crop following wheat harvest is adapted to all of Ohio and has been tried by a few growers in southern Michigan. The northern limit of this practice will be determined by the length of growing season. Successful wheat-soybean doublecropping requires quite specific management practices. These include:

(1)   Harvest wheat early. Wheat harvested at 20 to 25 percent moisture does not reduce yield and can advance the harvest date 5 to 7 days.

(2)   Remove straw, leave stubble high.

(3)   Select an early soybean variety; plant with no-tillage in narrow rows. Soybean plants do not grow large so narrow rows are needed to insure full canopy development. 'Grant' and 'Steele' cultivars have been good for Ohio conditions but are considered too early for single cropping.

(4)   Spray stubble with an appropriate herbicide.

To minimize chances of failure:

(5)   Planting should be stopped by July 10. Yield reductions of 1/3 bushel per day of planting delay are taking place at this time.

(6)   If the soil is dry at wheat harvest time, do not plant.

With good management, soybeans as a second crop have averaged 1200 kg/ha in northern Ohio. No-tillage seems to be the best choice system on a range of soils for this double cropping.

Silage Corn Following Hay Harvest

Alfalfa meadows are maintained until stands deteriorate and productivity declines. The meadow is destroyed and followed by corn plus one or more annual crops before being reestablished. Commonly, the meadow is plowed in the fall or early spring and no harvest is made prior to planting corn. Alfalfa represents an important source of high protein, high quality forage for dairy farmers, and the first harvest is usually the most productive.

Decreases in corn grain production with delayed planting are well documented by studies in Michigan and other states in this region. Some early work in Ohio, however, indicates that total corn plant yield is not affected in the same manner. As grain yield declines, corn plants tend to grow taller and compensate to some extent for losses in grain production. Total plant yields may peak a few days later than grain yields. Making a meadow harvest during the week of May 25 and planting corn into the stubble for a silage crop should produce an almost full corn yield with the hay as a bonus. Admittedly, the corn grain/stover ratio is moving in an unfavorable direction at this time, but perhaps not enough to impair quality for dairy animals.

As with wheat-soybean doublecropping, no-tillage planting conserves soil moisture and helps facilitate rapid establishment of the second crop. There is some risk involved in this system. The meadow crop extracts soil moisture and corn yields might be reduced in an extremely dry year. The soil does not require tillage to insure nitrogen release from the killed alfalfa crop (Triplett et al., 1979).

SOILS AND TILLAGE SELECTION

Scientists in Michigan, Ohio, and Indiana recognized that soil characteristics have a profound influence on crop response to tillage systems and developed lists of soils with suggestions for tillage selection or indications of soil suitability for no-tillage. In general, as drainage improves, the need for tillage required to maintain yields decreases. Tillage systems represent a continuum and some intensity of tillage should be optimum for any specific set of soil conditions. To date, this area is incompletely defined. This section will address two questions:

(1)   Will soil need to be tilled for some reason after 3 or 5 or 10 years of no-tillage production? Numerous people have speculated in print that this will be the case.

(2)   Continuous corn yields are reduced with no-tillage on poorly drained soils such as Hoytville and Brookston. Are there management strategies that will permit tillage reduction while maintaining yields on these soils?

Corn yields from crop rotation experiments located on two different soils are shown in Table 1 (Van Doren et al., 1976). The Wooster site is a better drained soil with a silt loam texture (Typic Fragiudalf) and was in meadow for 6 years prior to initiation of the study. This experiment is still being maintained and in recent years yields follow the same patterns shown here with no-tillage significantly greater than for plowed treatments. During individual years, yields vary with amount and distribution of rainfall during the growing season. During years with drought stress, no-tillage treatment yields are greater than for plowed treatments. With ample, well distributed rainfall, yields of plowed treatments are nearly equal to no-tillage. The site of this experiment is on a 3 to 4 percent slope and obvious erosion has taken place on plowed treatments during the course of the study. Besides tillage response, there is also a crop rotation effect. Untilled Wooster soil requires more than 60 percent of the soil surface covered with mulch for satisfactory no-tillage production. The mulch helps improve rainfall infiltration and reduces evaporation. This mulch is supplied by crop residue. With the yield trend, noted above, moldboard plowing for this site and soil becomes a yield reducing practice. After two decades, there is no evidence that this site will ever need tillage to maintain crop productivity.

In contrast, with no-tillage, continuous corn yields on the Hoytville site were significantly lower and almost equivalent to fall moldboard plowing for the two crop rotations (Table 1). Thus, no-tillage becomes a yield reducing practice for corn grown in monoculture. Hoytville soil is poorly drained, cracks when dry, has a silty clay loam surface texture, and is classified as a Mollic Ochraqualf. When yield trends with no-tillage became apparent, other

Table 1. Average corn yields for two locations, three rotations and two sets of tillage treatments, years 7-12. (Adapted from Van Doren et al., 1976).

Site	Rotation	No-till (kg/ha)	Plow (kg/ha)	Prob*
Wooster	Continuous corn	9400	8420	<0.001
Silt Loam	Corn-soybeans	9480	8720	0.03
	Corn-oats-hay	10450	9720	0.01
Hoytville	Continuous corn	6820	8000	<0.001
Silty Clay Loam	Corn-soybeans	7920	8260	ND
	Corn-oats-hay	8180	8390	ND

* Probability level at which yield differences between pairs of tillage treatments is equal to zero. ND means no difference (p >0.20).

studies were initiated to explore the questions of how reduced tillage systems might be adapted to poorly drained soils. While fall or winter plowing produces high yields, soil is exposed to wind and water erosion. Moldboard plowing these wet soils at planting time reduces yields approximately 15 percent.

Results from two studies that were contiguous to the rotation experiment described above provide information useful in formulating management strategies. Yields for individual years are shown for each experiment for the period 1968-1972 (Table 2). Experiment I was a date-of-planting study with continuous corn in which no tillage followed the previous season's no-tillage during one year and no-tillage followed corn grown in fall plowed soil the other three seasons. Experiment II represents the crop rotation study with no rotation of tillage. Experiment III had a factorial design that included previous crop, tillage, timing of tillage, and residue levels. Tillage was rotated in this experiment for all cropping sequences (Triplett and Van Doren, unpublished).

Table 2. Annual corn grain yields for various tillage and cropping history combinations, Experiments I, II and III.

Experiment I

| Year | Number Dates of Planting | Continuous Corn | | |
		Fall Plow (kg/ha)	No Till (kg/ha)	Difference[a] %
1968	15	7300	7000[b]	- 4
1969	17	6700	5400[c]	-20[d]
1970	14	5800	5200[b]	-10[d]
1971	17	9300	8700[b]	- 7[d]
Mean		7300	6600	-10[d]

Experiment II

| Year | Number Date of Planting | Continuous Corn | | | Corn-Oats-Hay | | |
		Fall Plow kg/ha	No Till kg/ha	Diff[a] %	Fall Plow kg/ha	No Till kg/ha	Diff[a] %
1968	15	5900	4900[c]	-17[d]	5900	6300[c]	+ 7[d]
1969	17	5900	4700[c]	-20[d]	6000	6000[c]	0
1970	14	7400	6200[c]	-16[d]	7600	6700[c]	-12[d]
1971	17	10500	9300[c]	-11[d]	11000	10800[c]	- 2
Mean		7400	6300[c]	-16[d]	7600	7400[c]	- 2

Experiment III

| Year | Number Date of Planting | Continuous Corn | | | Corn-Oats-Hay | | |
		Fall Plow kg/ha	No Till kg/ha	Diff[a] %	Fall Plow kg/ha	No Till kg/ha	Diff[a] %
1968	15	7800	7500[b]	- 4	8200	7800[b]	- 5
1969	17	7500	7100[b]	- 5	8400	8500[b]	+ 1
1970	14	7400	7300[b]	- 1	7400	7900[b]	+ 7
1971	17	10200	9900[b]	- 3	11400	10100[b]	-11[d]
Mean		8200	7900[b]	- 3	8800	8600[b]	- 2

[a]((No-till) - (Fall plow))/Fall plow
[b]Tillage rotated   [c]Tillage not rotated
[d]Difference significant < 5% level of probability

Yields for continuous no-tillage corn averaged 16 percent less than fall plowing in Experiment II during the 4 years shown (Table 2). In contrast, yields of continuous no-tillage corn in Experiment III averaged a non-significant 3 percent less than fall plowing. Plowing for the previous year's crop essentially eliminated the yield depression associated with no-tillage corn in monoculture. In Experiment I, tillage was rotated in 3 of 4 years. In 1969, no-tillage followed no-tillage and the percentage yield reduction was the same as for Experiment II. For two of the other three years with tillage rotated, no-tillage yields were significantly lower than for fall plowing. The percentage decrease, however, was not as great as for Experiment II where tillage was not rotated. For Experiments II and III, no-tillage yield was equal to fall plowing where corn followed some other crop. In this case, rotation of tillage was not a factor. Based on results from these studies, corn yields with no-tillage can be equal to corn yields with fall plowing on this poorly drained soil provided: (1) corn follows some other crop, regardless of prior tillage practice; and (2) for continuous corn, tillage is rotated.

Experiment III also contained a disking in the spring treatment and a spring moldboard plow treatment. Spring moldboard plowing had greatest yield reduction because of delayed planting and poor stands. Spring disking was no better than no-tillage when tillage was rotated. Thus, the producer on the poorly drained soils who does not fall plow should consider no-tillage or an intermediate system rather than moldboard plowing at planting time.

A major difference between the Wooster and Hoytville soils is the degree of drainage. On poorly drained soils, as indicated above, tillage may be required to maintain crop yields under some circumstances. An experiment was initiated on Hoytville soil to test the hypothesis that drainage and tillage systems are interrelated in crop yield response and that some intensity of tillage less than moldboard plowing may be adequate to maintain crop yields. This study included tile plus surface drainage or surface drainage alone as whole blocks. Soil was ridged as a means to further improve drainage and crops were planted on the ridge or on unridged soil. Besides fall moldboard plowing and no-tillage, fall chisel plowing was introduced as an intermediate system. Corn was grown in monoculture or rotated with soybeans. Results from the first year of this study after this crop and tillage history was established are shown in Table 3.

Table 3. Corn yield, drainage, tillage and rotation experiment, Hoytville silty clay loam soil, 1980.

System	Fall Moldboard Plow (kg/ha)	No-Tillage (kg/ha)	Chisel Plow (kg/ha)
+ Tile, rotation, + ridge	9550	9460	8600
+ Tile, rotation, - ridge	9550	9260	9550
+ Tile, cont. corn, + ridge	9260	9260	7120
- Tile, rotation, + ridge	9550	9160	8190
+ Tile, cont. corn, - ridge	9600	7800	8600
- Tile, rotation, - ridge	8770	8380	8970
- Tile, cont. corn, - ridge	8380	6630	7310
*LSD$_{0.05}$		1170	

*Least Significant Difference, 0.05 level of probability

Corn yields from the various tillage, drainage and rotation treatments varied but fell into three distinct groups. Each of the groups (Table 3) were significantly different from the others. Rotation and tillage effects present for experiments described earlier on the poorly drained soil were present in this study. However, for tile-drained soil, improving drainage further by planting on ridges maintained yields of continuous no-tillage corn at the same level as fall plowing. Corn yields for chisel plowing, as an intermediate system, were no better than no-tillage and influenced by the same management factors. The greatest yield reductions for no-tillage and chisel plowing were for continuous corn without tile or ridges. In this study planting on ridges partially substituted for tile drainage. This study is being continued to obtain a reasonable sampling of years.

Results from Ohio tillage studies indicated that tillage systems should be matched to soil characteristics. For better drained soils, reduced or no-tillage systems are the best choice for a variety of reasons. For poorly drained soils, reduced tillage systems may be used without yield reductions provided correct management is followed. These management factors include improved drainage, planting on ridges, rotation of crops, and rotation of tillage.

SUMMARY

Research on reduced and no-tillage crop production was initiated in Ohio (for the modern era) in 1960. Much of the early research effort focused on development of management systems and their evaluation for various crops, soils and locations. Topics addressed included: planting, fertility, control of weeds and other pests, soils and tillage response, mulch cover effects, soil erosion and pesticide movement, long term effects, and multiple cropping systems. This paper deals with soils and tillage selections and multiple cropping. Reduced tillage systems can become tools in development of new management strategies such as multiple cropping systems. Multiple cropping that may be applicable both to Ohio and parts of Michigan includes soybeans following wheat harvest and corn for silage following a first meadow harvest. Reduced and no-tillage practices are better adapted to well drained soils. On some well drained soils, no-tillage yields have been maintained for more than 20 years and show no indication of declining. On these soils, moldboard plowing reduces corn yields. On poorly drained soils, continuous no-tillage corn reduces yields 15 to 20 percent compared to fall plowing. On these same soils with improved drainage, rotation of crops, or rotation of tillage (alternating plowing and no-tillage), yields are maintained approximately equal to fall plowing.

LITERATURE CITED

Faulkner, E.H. 1943. Plowman's Folly. University of Oklahoma Press, Norman, OK.

Free, G.R., S.N. Festig, and C.E. Bay. 1963. Zero tillage for corn following sod. Agron. J. 55: 207-208.

Triplett, G.B., Jr., F. Haghiri, and D.M. Van Doren Jr. 1979. Plowing effects on corn yield response to nitrogen following alfalfa. Agron. J. 71:801-803.

Van Doren, D.M., Jr., G.B. Triplett Jr. and J.E. Henry. 1976. Influence of long term tillage, crop rotation and soil type combinations on corn yield. Soil Sci. Soc. Am. J. 40:100-105.

CHAPTER 7

CONSERVATION TILLAGE IN MICHIGAN: A STATUS REPORT

Dwight Quisenberry
Soil Conservation Service
U.S. Department of Agriculture
East Lansing, Michigan 48823

Conservation tillage continues to gain the attention of landowners as a cost-effective way to reduce wind and water erosion.

According to a survey which was completed by Soil Conservation Service (SCS) field office employees in Michigan, there were approximately 63,294 ha of no-till corn and 12,037 ha of no-till soybeans grown in Michigan during 1984 representing a 50 percent increase over 1983 when approximately 42,523 ha of no-till corn and 8,100 ha of no-till soybeans were grown. The 1984 increase indicates expanded acceptance and use of no-till. Similar gains were recorded for no-till wheat which increased from 4,455 to 8,505 ha and hay and pasture seedings from 2,430 to 6,131 ha during the same period (Table 1). Other no-till crops grown were barley, oats, sweet corn, rye, asparagus, popcorn, tomatoes, flax and buckwheat. More detailed county crop data are presented in Tables 2, 3 and 4 and Figures 1 and 2.

Increases were due to promotional efforts by the Michigan Association of Conservation Districts, local soil conservation districts, technical and educational work by the Soil Conservation Service and Michigan State University Cooperative Extension Service (see Table 5). The interest of farmers in saving soil and money as well as the increasing support of news media and private industry are also cited as reasons for the increases in no-till farming.

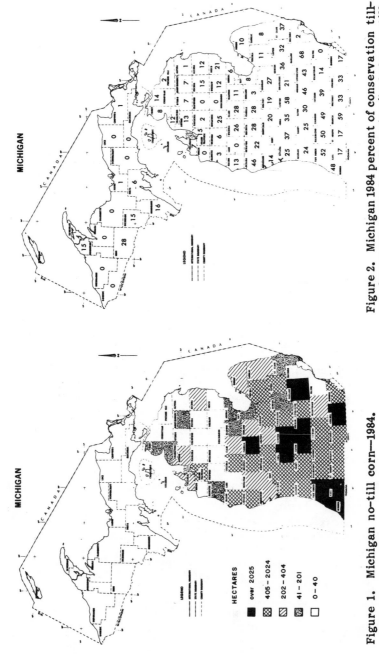

Figure 2. Michigan 1984 percent of conservation tillage for row crops and small grains (includes no-till, ridge till, strip till, mulch till and reduced till).

Figure 1. Michigan no-till corn—1984.

Table 1. Summary of no-till plantings in Michigan, 1980-1984 (hectares).

Year	Corn	Soybeans	Small Grains	Pasture and Hay
1980	20,250	2,025	1,620	1,215
1981	34,020	3,240	2,430	1,215
1982	45,360	5,670	4,050	1,620
1983	42,523	8,100	4,050	2,430
1984	63,294	12,037	8,505	5,670

The following are typical suggestions to farmers trying the no-till planting of corn or soybeans for the first time: (1) select a field with corn or soybean residue on the surface, good drainage and no unusual weed or soil compaction problems; (2) apply lime and fertilizer needed according to soil tests; (3) plant crops when they should be planted using a properly adjusted no-till planter; (4) use proper herbicides and insecticides; and (5) monitor planting, seed germination, seedling emergence, plant growth and crop production to identify needed system adjustments and explain crop differences.

There were 157 entrants from 33 counties that participated in the 1984 Ortho No-Till Corn Yield Contest in Michigan. Yields ranged from 61 to 178 bushels per acre with an average yield of 120 bushels.

Nearly half of the state's cropland needs conservation treatment to reduce soil erosion to tolerable levels and maintain the soil resource base. Much of the state's cropland is farmed more intensively than a decade or two ago. Corn and soybean production has increased significantly in the last ten years, and over half of the state's cropland is now used to grow these and other row crops. As a result of this intensification of farming, wind and water erosion are urgent problems.

No-till and other forms of conservation tillage are making gains in the state. The total no-till acreage is still small but is expected to increase significantly in the next few years.

Table 2. Michigan conservation tillage survey for corn, wheat, pasture and hay, soybeans, and selected crops for 1984 (hectares).

SCS* Area and Field Offices	Corn	Wheat	Pasture and Hay	Soybeans	Other**	Ridge Till	Mulch Till
MARQUETTE—1							
Crystal Falls	—	—	—	—	—	—	162
Escanaba	25	—	243	—	—	—	445
Hancock	—	—	—	—	—	—	445
Iron Mountain	—	—	—	—	—	—	425
Manistique	—	—	48	—	—	—	—
Marquette	8	—	—	—	—	—	—
Ontonagon	—	—	—	—	—	—	—
Sault Ste. Marie	—	20	—	—	—	—	—
Stephenson	20	—	44	—	—	—	2,227
Total	53	20	336	—	—	—	3,705
TRAVERSE CITY—2							
Bellaire	425	151	—	—	—	—	255
Big Rapids	481	61	260	—	50	—	4,815
Boyne City	222	—	29	—	—	—	364
Cadillac	81	—	—	—	—	—	121
Fremont	202	8	24	24	303	—	3,973
Lake City	—	—	8	—	—	—	1,903
Lake Leelanau	202	—	73	—	38	—	405
Onekama	56	—	—	—	—	—	66
Reed City	405	40	81	18	4	—	2,162

Table 2. Continued.

SCS* Area and Field Offices	Corn	Wheat	Pasture and Hay	Soybeans	Other**	Ridge Till	Mulch Till
TRAVERSE CITY (Cont.)							
Scottville	486	—	8	—	255	—	1,215
Shelby	743	60	40	—	3,645	—	3,321
Traverse City	405	—	10	—	—	—	1,215
Total	3,712	323	534	42	4,297	—	19,819
GRAYLING—3							
Alpena	111	—	—	—	—	—	40
Bay City	81	—	—	36	—	56	4,657
Cheboygan	—	—	—	—	—	—	567
East Tawas	—	—	12	—	—	—	—
Gaylord	18	—	—	—	—	—	157
Gladwin	66	4	24	—	—	—	1,202
Harrison	283	—	32	—	—	—	1,944
Harrisville	265	—	—	—	—	—	170
Ithaca	216	415	8	73	—	402	16,200
Midland	362	271	9	119	—	—	37
Mt. Pleasant	4,252	60	40	486	11	—	9,173
Rogers City	—	—	14	—	—	—	141
Saginaw	1,563	—	—	637	—	—	26,163
Standish	162	—	—	—	—	6	1,215
West Branch	81	6	93	—	—	—	1,559
Total	7,464	758	234	1,353	11	464	63,230

Table 2. Continued.

SCS* Area and Field Offices	Corn	Wheat	Pasture and Hay	Soybeans	Other**	Ridge Till	Mulch Till
GRAND RAPIDS—4							
Allegan	1,599	16	56	101	—	—	13,383
Cassopolis	2,025	121	202	445	8	—	7,492
Centreville	1,417	—	54	486	—	—	8,910
Charlotte	4,050	405	162	202	—	—	15,673
Coldwater	1,215	2,106	—	81	48	344	810
Grand Haven	810	—	—	—	—	—	8,505
Grand Rapids	1,831	166	130	18	—	324	13,672
Hastings	814	83	60	30	2	—	7,925
Ionia	3,685	202	206	405	—	182	19,642
Kalamazoo	1,822	121	—	60	—	—	20,947
Marshall	2,004	202	62	299	23	—	23,895
Muskegon	202	—	52	8	—	—	2,025
Paw Paw	4,374	121	141	364	5,627	—	14,001
St. Johns	6,358	2,673	68	3,888	—	473	30,496
St. Joseph	3,240	283	162	1,215	4	405	13,587
Stanton	1,579	81	70	32	—	—	9,882
Total	37,030	6,571	1,432	7,638	5,714	1,729	210,849
FLINT—5							
Adrian	2,430	121	40	486	—	607	34,425
Ann Arbor	259	—	32	89	—	170	7,897
Bad Axe	303	—	—	48	—	—	12,150

Table 2. Continued.

SCS* Area and Field Offices	Corn	Wheat	Pasture and Hay	Soybeans	Other**	Ridge Till	Mulch Till
FLINT (Cont.)							
Caro	648	6	—	202	—	178	12,453
Hillsdale	931	162	202	405	—	162	34,830
Howell	810	—	40	121	—	—	9,760
Jackson	1,944	121	202	162	3	—	18,751
Lapeer	729	—	40	81	—	81	13,162
Mason	2,328	60	64	955	—	81	27,074
Monroe	1,012	50	10	607	—	121	11,947
Mt. Morris	141	—	—	101	—	810	11,399
Owosso	1,113	234	48	101	—	24	12,028
Pontiac	2,025	—	372	32	16	168	3,783
Port Huron	97	—	—	16	—	648	17,010
Richmond	—	—	—	5	—	243	121
Sandusky	259	—	—	—	—	319	8,262
Total	15,033	757	3,592	3,004	19	3,614	235,057
State Total	63,293	8,442	6,130	12,037	10,043	5,808	532,662

* SCS = Soil Conservation Service

** Other no-till includes: barley, oats, sorghum, grass, field beans, sweet corn, string beans, rye, asparagus, popcorn and vegetables.

Table 3. Michigan conservation tillage survey for selected crops in 1984.

SCS Area and Field Offices	No-Till ------- Hectares -------							
	Oats	Rye	Barley	Flax	Buck-Wheat	Sweet Corn	Pop-Corn	Tomatoes
MARQUETTE—1								
Crystal Falls	—	—	—	—	—	—	—	—
Escanaba	—	—	—	—	—	—	—	—
Hancock	—	—	—	—	—	—	—	—
Iron Mountain	—	—	—	—	—	—	—	—
Manistique	—	—	—	—	—	—	—	—
Marquette	—	—	—	—	—	—	—	—
Ontonagon	—	—	—	—	—	—	—	—
Sault Ste. Marie	4	—	—	4	—	—	—	—
Stephenson	—	—	—	—	12	—	12	—
Total	4	—	—	4	12	—	12	—
TRAVERSE CITY—2								
Bellaire	—	—	—	—	—	—	—	—
Big Rapids	—	101	—	—	—	—	—	—
Boyne City	—	—	—	—	—	—	—	—
Cadillac	—	—	—	—	—	—	—	—
Fremont	—	—	—	—	—	—	—	—
Lake City	—	—	—	—	—	—	—	—
Lake Leelanau	—	—	—	—	—	—	—	—
Onekama	—	—	—	—	—	—	—	—
Reed City	—	101	—	—	—	—	—	—

Table 3. Continued.

SCS Area and Field Offices	Oats	Rye	Barley	Flax	Buck-Wheat	Sweet Corn	Pop-Corn	Tomatoes
TRAVERSE CITY (Cont.)								
Scottville	—	—	—	—	—	—	—	—
Shelby	—	41	—	—	—	—	—	—
Traverse City	—	—	—	—	—	—	—	—
Total	—	243	—	—	—	—	—	—
GRAYLING--3								
Alpena	—	—	—	—	—	—	—	—
Bay City	—	—	—	—	—	—	—	—
Cheboygan	—	—	—	—	—	—	—	—
East Tawas	—	—	—	—	—	—	—	—
Gaylord	—	—	—	—	—	—	—	—
Gladwin	—	—	—	—	—	—	—	—
Harrison	—	—	—	—	—	—	—	—
Harrisville	—	—	8	—	—	—	—	—
Ithaca	—	—	—	—	—	—	—	—
Midland	—	—	—	—	—	—	—	—
Mt. Pleasant	28	—	—	—	—	—	—	—
Rogers City	—	—	5	—	—	—	—	—
Roscommon	—	—	—	—	—	—	—	—
Standish	—	—	—	—	—	—	—	—
Saginaw	—	—	—	—	—	—	—	—
West Branch	2	—	—	—	—	—	—	—
Total	30	—	13	—	—	—	—	—

Table 3. Continued.

SCS Area and Field Offices	No-Till — Hectares							
	Oats	Rye	Barley	Flax	Buck-Wheat	Sweet Corn	Pop-Corn	Tomatoes
GRAND RAPIDS--4								
Allegan	—	16	—	—	—	—	—	—
Cassopolis	—	—	—	—	—	—	—	—
Centreville	—	—	40	—	—	—	—	—
Charlotte	202	40	—	—	—	—	48	—
Coldwater	—	1,215	—	—	—	—	—	—
Grand Haven	49	48	7	—	—	—	—	—
Grand Rapids	5	—	—	—	—	2	—	0.4
Hastings	—	—	—	—	—	—	—	—
Ionia	—	—	—	—	—	—	—	—
Kalamazoo	—	—	—	—	—	—	—	—
Marshall	—	—	—	—	—	—	—	—
Muskegon	—	—	—	—	—	—	—	—
Paw Paw	—	81	—	—	—	—	—	—
St. Johns	—	20	—	—	—	—	—	—
St. Joseph	40	—	—	—	—	—	—	4
Stanton	—	—	—	—	—	—	—	—
Total	294	1,421	47	—	—	2	48	4.4
FLINT--5								
Adrian	—	—	—	—	—	—	—	—
Ann Arbor	—	—	—	—	—	—	—	—
Bad Axe	—	—	—	—	—	—	—	—
Caro	—	—	—	—	—	—	—	—

Table 3. Continued.

SCS Area and Field Offices	No-Till — — — — Hectares — — — —							
	Oats	Rye	Barley	Flax	Buck-Wheat	Sweet Corn	Pop-Corn	Tomatoes
FLINT (Cont.)								
Hillsdale	24	4	—	—	—	—	—	—
Howell	—	—	—	—	—	—	—	—
Jackson	28	—	3	—	—	—	—	—
Lapeer	—	—	—	—	—	—	—	—
Mason	—	—	—	—	—	—	—	—
Monroe	12	—	—	—	—	—	—	—
Mt. Morris	—	—	—	—	—	—	—	—
Owosso	16	32	—	—	—	—	—	—
Pontiac	—	—	—	—	—	—	—	—
Port Huron	—	—	—	—	—	—	—	—
Richmond	—	—	—	—	—	—	—	—
Sandusky	—	—	—	—	—	—	—	—
Total	81	36	3	—	—	—	—	—
State Total	409	1,700	63	4	12	2	48	4.4

Table 4.   Percent of conservation tillage for Michigan in 1984.

SCS* Area and Field Offices	Acres Planted**	Acres of Conservation Tillage***	Percent of Conservation Tillage
MARQUETTE--1			
Crystal Falls	587	162	28
Escanaba	4,276	268	6
Hancock	1,061	162	15
Iron Mountain	2,875	425	15
Manistique	2,184	--	--
Marquette	712	8	1
Ontonagon	275	--	--
Sault Ste. Marie	5,211	20	1
Stephenson	13,770	2,247	16
Total	30,954	3,293	10
TRAVERSE CITY--2			
Bellaire	8,565	832	10
Big Rapids	18,338	5,150	28
Boyne City	6,075	587	10
Cadillac	3,551	202	6
Fremont	19,359	4,183	22
Lake City	7,654	1,903	25
Lake Leelanau	9,001	645	7
Onekama	3,602	123	3
Reed City	9,219	2,428	26
Scottville	9,697	1,308	13
Shelby	15,390	7,102	46
Traverse City	10,997	1,620	15
Total	121,453	26,088	21
GRAYLING--3			
Alpena	8,960	151	2
Bay City	62,775	4,722	8
Cheboygan	4,171	567	14
East Tawas	5,766	1,215	21
Gaylord	2,444	176	7
Gladwin	11,082	1,212	11
Harrison	7,978	2,227	28
Harrisville	2,845	435	12
Ithaca	90,720	17,308	19

Table 4. Continued.

SCS* Area and Field Offices	Acres Planted**	Acres of Conservation Tillage***	Percent of Conservation Tillage
GRAYLING (Cont.)			
Midland	24,502	790	3
Mt. Pleasant	50,017	13,984	28
Rogers City	8,176	141	2
Roscommon	42	--	--
Standish	24,101	1,383	6
Saginaw	104,355	28,364	27
West Branch	9,699	1,646	17
Total	417,640	74,327	18
GRAND RAPIDS--4			
Allegan	63,299	14,889	24
Cassopolis	60,770	10,092	17
Centreville	62,552	10,813	17
Charlotte	67,230	20,331	30
Coldwater	77,824	4,604	6
Grand Haven	37,017	9,315	25
Grand Rapids	43,287	16,182	37
Hastings	35,595	8,970	25
Ionia	68,850	24,252	35
Kalamazoo	45,932	14,851	50
Marshall	54,027	26,425	49
Muskegon	15,633	2,235	14
Paw Paw	46,546	24,366	52
St. Johns	73,038	42,593	58
St. Joseph	38,475	18,310	48
Stanton	53,626	10,602	20
Total	770,804	266,939	35
FLINT--5			
Adrian	114,210	38,070	33
Ann Arbor	57,024	8,051	14
Bad Axe	129,600	12,502	10
Caro	149,465	13,488	11
Hillsdale	109,435	36,490	33
Howell	24,583	10,692	43
Jackson	53,331	20,982	39
Lapeer	43,497	14,053	32

Table 4. Continued.

SCS* Area and Field Offices	Acres Planted**	Acres of Conservation Tillage***	Percent of Conservation Tillage
FLINT (Cont.)			
Mason	66,484	30,500	46
Monroe	81,708	13,557	17
Mt. Morris	34,501	12,452	36
Owosso	65,002	13,502	21
Pontiac	9,015	6,126	68
Port Huron	47,790	17,771	37
Richmond	19,113	370	2
Sandusky	104,390	8,841	8
Total	1,084,854	257,452	24
State Total	2,425,707	628,101	26

*   CSC = Soil Conservation Service

** Crops included are: corn, small grain, soybeans, grain sorghum, vegetable crops and other row crops. These factors and figures are taken from information submitted to the Conservation Tillage Information Center.

*** Conservation tillage includes: no-till, ridge till, strip till, mulch till, and reduced till.

Table 5. Summary of no-till equipment and conservation tillage meetings in Michigan in 1984.

SCS Area and Field Offices	Corn Planters			Soybean Planters			Hay/Pasture Drills			Mtgs.**
	Mfg.*	No. Rent, Ind. Lease Farmer or Free	No. Use Only	Mfg.*	No. Rent, Ind. Lease Farmer or Free	No. Use Only	Mfg.*	No. Rent, Ind. Lease Farmer or Free	No. Use Only	No. Held
MARQUETTE—1										
Crystal Falls	—	—	—	—	—	—	4	1	—	—
Escanaba	1	—	6	—	—	—	1	—	1	1
Hancock	—	—	—	—	—	—	4	1	—	—
Iron Mountain	—	—	—	—	—	—	4	1	—	—
Manistique	—	—	—	—	—	—	—	—	—	2
Marquette	6	—	1	—	—	—	10	1	—	—
Ontonagon	—	—	—	—	—	—	—	—	—	—
Sault Ste.Marie	—	—	—	—	—	—	1	1	—	—
Stephenson	1,2,6	—	9	—	—	—	1,4	2	—	2
Total		0	16		0	0		7	1	5
TRAVERSE CITY—2										
Bellaire	1,5	—	14	—	—	—	3	—	1	2
Big Rapids	1,6	—	20	3	1	—	3	1	—	—
Boyne City	1,2,6	3	4	—	—	—	10	2	—	3
Cadillac	6	1	—	—	—	—	—	—	—	2
Fremont	2,5,6	3	42	6	—	1	2,3	2	1	1
Lake City	—	—	—	—	—	—	—	—	—	3
Lake Leelanau	1,2,6	—	5	—	—	—	3	1	—	—

Table 5. Continued.

SCS Area and Field Offices	Corn Planters Mfg.*	No. Rent, Ind. Lease Farmer or Free	No. Use Only	Soybean Planters Mfg.*	No. Rent, Ind. Lease Farmer or Free	No. Use Only	Hay/Pasture Drills Mfg.*	No. Rent, Ind. Lease Farmer or Free	No. Use Only	Mtgs.** No. Held
TRAVERSE CITY (Cont.)										
Onekama	1,2	—	2	—	—	—	—	—	—	—
Reed City	1,2,6	6	7	—	—	—	1,3,4	3	—	3
Scottville	1,2,6	1	16	—	—	—	10	1	1	1
Shelby	1,2,5,6	3	25	—	—	—	—	1	1	2
Traverse City	1	1	8	—	1	—	1	—	—	—
Total	—	18	135	—	1	1	—	11	4	17
GRAYLING—3										
Alpena	6	—	3	—	—	—	—	—	—	—
Bay City	1,2,6	1	8	6	—	1	1	—	1	—
Cheboygan	—	—	—	—	—	—	—	—	—	—
East Tawas	1,2,5	2	2	1,2,5	2	1	3,4	1	1	2
Gaylord	1	1	1	1	—	—	4	1	—	—
Gladwin	1,2	1	4	1	1	—	3,4	2	—	2
Harrison	1	—	6	1	—	6	3	2	—	—
Harrisville	1	—	2	—	—	—	1,4	1	1	2
Ithaca	1,2,5,6	2	8	1,2,5,6	1	8	—	—	1	1
Midland	1,2,5	1	6	1,2,5	1	6	—	—	—	1
Mt. Pleasant	1,2,5,6	—	61	—	—	—	3	1	—	1

Table 5. Continued.

SCS Area and Field Offices	Corn Planters			Soybean Planters			Hay/Pasture Drills			Mtgs.**
	Mfg.*	No. Rent, Ind. Lease Farmer or Free	No. Use Only	Mfg.*	No. Rent, Ind. Lease Farmer or Free	No. Use Only	Mfg.*	No. Rent, Ind. Lease Farmer or Free	No. Use Only	No. Held
GRAYLING (Cont.)										
Rogers City	6	—	1	—	—	—	10	—	1	—
Roscommon	—	—	—	—	—	—	3	—	1	—
Standish	1,5,6	5	20	1,5,6,9,18	6	21	9	1	—	5
Saginaw	—	—	3	—	—	—	—	1	—	—
West Branch	2,17	—	2	—	—	—	3	1	5	1
Total		12	154		11	43		10	5	15
GRAND RAPIDS—4										
Allegan	1,2,6	3	33	9,12	2	—	4,9,12	3	—	2
Cassopolis	1,2,6,11	1	29	9	1	—	9	1	—	1
Centreville	1,2,5,6	1	40	1,2,5,6	1	40	4,12	1	—	—
Charlotte	1,2,5,6	2	69	1,2,5,6,7,9	3	69	7,9,10	3	—	7
Coldwater	1,5,6	1	35	3,9	—	2	3,9	—	2	3
Grand Haven	1,3,6	6	24	—	—	—	—	—	—	2
Grand Rapids	1,2,5,6,13	4	33	7,12,15	3	1	7,12,15	3	1	2
Hastings	1,6	4	15	1,7,10	2	2	7,10	2	—	1
Ionia	1,2,5,6,13	8	42	1,5,7,15	5	—	4,7,15	3	—	6
Kalamazoo	1,2,5,6	1	12	9	1	—	9	1	—	3
Marshall	1,2,5,6,13,14,17	4	16	1,5,6,9	4	2	9,10	3	—	2

Table 5. Continued.

SCS Area and Field Offices	Corn Planters			Soybean Planters			Hay/Pasture Drills			Mfgs.**
	Mfg.*	No. Rent, Ind. Lease Farmer or Free	No. Use Only	Mfg.*	No. Rent, Ind. Lease Farmer or Free	No. Use Only	Mfg.*	No. Rent, Ind. Lease Farmer or Free	No. Use Only	No. Held
GRAND RAPIDS (Cont.)										
Muskegon	1,2,5,6	1	20	1	--	1	10,12	2	--	--
Paw Paw	1,2,5,6,17	7	23	1,9	1	15	19	--	2	5
St. Johns	1,2,5,6,13,14	7	75	7,9,10	3	2	7,9,10	3	2	2
St. Joseph	1,2,5,6	5	60	1,2,5,6	5	60	9,10	2	1	2
Stanton	1,5,6	2	17	--	--	--	19	1	--	6
Total		57	543		31	194		28	8	44
FLINT--5										
Adrian	1,2,6	2	17	9,10	1	1	9,10	1	1	2
Ann Arbor	1,2,5,6	2	23	1,2,6	2	17	1	2	2	1
Bad Axe	1,2,5	3	9	1,2,5	3	9	--	--	--	1
Caro	1,5,6	3	18	5,6,9	1	4	9	1	--	8
Hillsdale	1,2,5	1	280	5,9	--	8	1,7,10	--	9	3
Howell	1,2,5,17	7	26	6,17	1	2	4,10	--	2	2
Jackson	1,2,5,6	1	40	1,9	1	2	9,10	1	1	4
Lapeer	1,2,5,17	3	37	--	--	--	--	--	--	2
Mason	1,2,5,6,17	3	27	1,2,5,6,16,17	4	28	1,3,8,16	4	3	2
Monroe	1,2,5,6	1	50	--	--	--	9,10	2	--	1

Table 5. Continued.

SCS Area and Field Offices	Corn Planters			Soybean Planters			Hay/Pasture Drills			Mtgs.**
	Mfg.*	No. Rent, Ind. Lease Farmer or Free	No. Use Only	Mfg.*	No. Rent, Ind. Lease Farmer or Free	No. Use Only	Mfg.*	No. Rent, Ind. Lease Farmer or Free	No. Use Only	No. Held
FLINT (Cont.)										
Mt. Morris	1,5,6	1	2	—	—	—	—	—	—	—
Owosso	1,2,5,6	3	17	1,2,6,9	3	15	9	1	1	2
Pontiac	—	—	—	10	1	—	10	1	—	2
Port Huron	1,5,17	2	3	—	—	—	—	—	—	—
Richmond	1,6	—	20	1,6	—	20	—	—	—	1
Sandusky	1,2	—	5	—	—	—	—	13	19	—
Total		32	574		17	106		69	37	31
State Total		119	1,422		60	344				112

*1. John Deere   6. International   11. Kinze   16. Haybuster
2. Allis Chalmers   7. Marliss   12. M & W   17. New Idea
3. Moore   8. Farnam   13. Hiniker   18. Great Plain
4. Zip Seeder   9. Lilliston   14. Buffalo   19. Brillion
5. White   10. Tye   15. Vermeer

**Meetings

CHAPTER 8

FERTILIZER MANAGEMENT FOR CONSERVATION TILLAGE

Maurice L. Vitosh, William H. Darlington,
Charles W. Rice and Donald R. Christenson
Department of Crop and Soil Sciences
Michigan State University
East Lansing, Michigan 48824

INTRODUCTION

Tillage practices can have a dramatic effect on the physical, chemical and biological properties of soils. Because these properties are interrelated, a change in a physical property of the soil often results in a change in chemical and biological properties and vice versa. Nutrient uptake and fertilizer efficiency are also greatly affected by different tillage systems. Many researchers have identified these soil property changes and have related them to nutrient uptake and fertilizer efficiency.

The objectives of this paper are: (1) To review the scientific literature with regards to tillage effects on nutrient uptake by corn and fertilizer efficiency; and (2) To summarize the information and its implications to the farmer.

LITERATURE REVIEW

The following terminology as described by the Soil Science Society of America (Anon., 1978) will be used in this paper:

Conventional tillage - the combined primary and secondary tillage operations normally performed in preparing a seedbed for a given geographical area.

Minimum tillage - the minimum soil manipulation necessary for crop production or meeting tillage requirements under existing soil and climatic conditions.

No-tillage - a crop production system whereby a crop is planted directly into a seedbed not tilled since harvest of the previous crop.

Reduced tillage - a tillage sequence in which the primary operation is performed in conjunction with planting procedures in order to reduce or eliminate secondary tillage operations.

Nitrogen

Nitrogen availability in no-tillage systems has recently received considerable attention. Thomas et al. (1973) observed nearly a 50 percent reduction in nitrate nitrogen in the surface 15 cm of soil under no-tillage. Much of this reduced availability of nitrogen has been attributed to lower rates of mineralization of organic matter (Phillips et al., 1980), increased nitrogen immobilization (Rice and Smith, 1984), and greater losses due to denitrification (Rice and Smith, 1982) and leaching (Thomas et al., 1973). Rice and Smith (1984) found that 13 to 24 percent of the nitrogen applied to the surface of three Kentucky soils was immobilized after 5 weeks. This was 1.5 to 2 times greater than the plowed soils.

Greater nitrogen losses due to leaching and denitrification in no-tillage systems have usually been attributed to higher water contents and greater infiltration of water. No-tillage fields generally average 15 to 30 percent more water in the surface 30 cm than conventionally tilled soils (Darlington, 1983). Surface residues reduce moisture loss by decreasing evaporation early in the growing season (Blevins et al., 1971). Surface residues also improve water infiltration (Blevins and Cook, 1970). Greater earthworm activity in undisturbed soils has also been found to increase infiltration and promote deeper percolation of water (Graff, 1969).

Tillage also indirectly influences nitrogen availability by altering the soil microbial population. Doran (1980a; 1980b) found that all groups of microorganisms were present in higher populations in the surface 7.5 cm of no-tillage soils. He attributed the increase to higher moisture and organic matter in the surface. Deeper in the soil profile aerobic populations were 25 to 51 percent lower in no-tillage soils compared to plowed soils. Facultative anaerobes and denitrifiers were found in greater numbers and represented a greater proportion of the total microbial population of no-tillage soils. Consequently the potential for mineralization and nitrification is greater in conventionally tilled soils while denitrification is greater in undisturbed soils. The high population of microbial organisms near the surface may also result in increased immobilization of surface applied nitrogen fertilizers.

Increased urease activity in surface residues also results in greater ammonia volatilization losses of surface applied urea fertilizers (Bandel et al., 1980; Fox and Hoffman, 1981; Terman, 1979; Touchton and Hargrove, 1982). Fox and Hoffman (1981) estimated that at least 35 percent of the surface applied urea in their studies was lost through ammonia volatilization.

The relative efficiency of nitrogen fertilizers in various tillage systems has been thoroughly studied. Lower nitrogen efficiencies have been found in no-tillage systems particularly when nitrogen rates were suboptimal (Bandel et al., 1975; Moschler and Martens, 1975; Triplett and Van Doren, 1969). Higher nitrogen rates have either resulted in equivalent nitrogen recoveries (Legg et al., 1979; Moschler and Shear, 1975; Moschler et al., 1972) or lower nitrogen recoveries for the conventional tillage systems as compared to no-tillage (Bandel et al., 1975; Moschler and Martens, 1975; Triplett and Van Doren, 1969).

The efficiencies of nitrogen uptake in many studies have been greatly influenced by the source of nitrogen fertilizer. Many researchers have used ammonium nitrate to avoid ammonia volatilization losses which can occur with urea fertilizers. Bandel et al. (1980) in Maryland, Moschler and Jones (1974) in Virginia and Fox and Hoffman (1981) in Pennsylvania all found variable results when comparing the effectiveness of urea, UAN (urea-ammonium nitrate solution) and ammonium nitrate. However, all concluded that the variability was due to the relationship between rainfall and time of application. Any rainfall which occurred shortly after application resulted in little or no difference between the sources of nitrogen. Fox and Hoffman (1981) observed the following relationship between rainfall and effectiveness of urea and UAN:

(1)   If 10 mm or more of rainfall fell within 48 hours after surface application, little or no ammonia volatilization occurred.

(2)   If only 3 to 5 mm of rain fell within 5 days after the urea was surface applied, volatilization losses were estimated at 10 to 30 percent.

(3)   If no rain occurred within 6 days of application ammonia volatilization losses from urea were estimated to be greater than 30 percent.

Nitrogen fertilizer placement can also greatly affect the efficiency of nitrogen uptake. Bandel et al. (1980), Mengel et al. (1982) and Touchton and Hargrove (1982) all concluded that banding or injection of nitrogen below the soil surface resulted in the greatest uptake of nitrogen. Touchton and Hargrove (1982) also found that surface banded UAN solution resulted in near maximum efficiency even though it was not injected. The greater efficiency of subsurface nitrogen fertilizer placement is generally attributed to reduced ammonia volatilization and nitrogen immobilization by surface residues.

Phosphorus and Potassium

Lower mobility of phosphorus and to a lesser extent potassium often results in the accumulation of these nutrients near the surface when fertilizers are surface broadcast in no-tillage systems (Ketcheson, 1980; Lal, 1979; Moschler and Martens, 1975). The stratification of these nutrients in the soil surface has been of much concern to scientists, particularly when no-tillage soils become dry. Plant analysis, however, generally shows comparable levels of phosphorus uptake between no-tillage and conventional tillage systems (Belcher and Ragland, 1972; Estes, 1972; Moschler and Martens, 1975). The effectiveness of surface applied phosphorus depends greatly on the initial level of soil phosphorus. When phosphorus soil tests are high, yield and phosphorus uptake are not significantly different (Belcher and Ragland, 1972; Tripplett and Van Doren, 1969). When phosphorus soil tests are low, banded phosphorus fertilizer is usually superior (Eckert, 1983; Johnson, 1983). Increased root activity near the soil surface in no-tillage soils is often cited as the reason for the availability of surface applied phosphorus fertilizers. The increased root activity is believed to be due to the higher soil moisture under the surface mulch of no-tillage soils. Phillips et al. (1971) reported 10 times more corn roots early in the growing season in the top 5.1 cm of soil in no-tillage plots compared to conventionally tilled.

Moschler and Martens (1975) also found greater residual phosphorus when no-tillage soils were repeatedly cropped in the greenhouse. Ketcheson (1980) reported higher levels of soil test phosphorus in no-tillage soils when averaged over 30 cm of depth and concluded that less total phosphorus fixation occurs in no-tillage soils.

Like phosphorus, potassium usually becomes stratified in no-tillage soils. Potassium, however, usually moves to a greater depth than phosphorus in no-tillage soils (Fink and Wesley, 1974; Triplett and Van Doren, 1969). Some researchers believe that this stratification increases the total amount of exchangeable potassium in the rooting zone (Hargrove et al., 1982; Ketcheson, 1980). A few studies, however, have reported reduced potassium uptake with no-tillage and minimum tillage systems (Ketcheson, 1980; Moncrief and Schulte, 1981). Moncrief and Schulte (1981) postulated that poor soil aeration in their no-tillage plots was responsible for the reduced potassium uptake.

The application of nitrogen, phosphorus and lime has also been shown to improve potassium uptake in no-tillage systems (Lal, 1979; Moncrief and Schulte, 1981)

Soil Acidity and Lime

Surface applications of acid forming nitrogen fertilizers often lowers the surface pH of no-tillage soils, particularly the sandy soils with low buffering capacities (Blevins et al., 1977; Fox and Hoffman, 1981; Moschler et al., 1973). Surface applied lime, however, has been very effective in controlling surface acidity in no-tillage systems. In the few studies recently conducted, there have been no adverse effects of surface applied limestone. Virginia studies (Moschler et al., 1973; Shear and Moschler, 1969) have shown greater yield benefits with corn from liming no-tillage than plowed soils.

Micronutrients

Few studies have looked at the effects of tillage on micronutrient uptake and redistribution in soils. Hargrove et al. (1982) found greater concentrations of zinc and manganese in the surface 7.5 cm of no-tillage soils compared to conventional tilled soils. Extractable copper was similar for all soil depths and tillage treatments. Estes (1972) found reduced levels of zinc, molybdenum, boron and aluminum in corn leaf tissue under no-tillage conditions. He concluded that surface liming had adversely affected the availability of zinc and boron. Manganese uptake is often influenced by nitrogen rates. Increasing nitrogen rates usually results in increased manganese uptake (Lal, 1979; Lutz and Lillard, 1973).

SUMMARY AND RECOMMENDATIONS

A review of the literature indicates that nitrogen fertilization and management are the most important soil fertility factors to consider when growing no-tillage corn. Reduced phosphorus availability in no-tillage or minimum tillage systems does not appear to be a major limiting factor for no-tillage corn production. If phosphorus soil test levels are low, banded phosphorus is required for optimum yields. The importance of banded phosphorus in no-tillage systems in Michigan may be even more important than demonstrated in other areas of the U.S. because of colder soils in the spring.

The availability of potassium in no-tillage soils also does not appear to be a major limiting factor in obtaining high yields of corn. On fine textured soils where soil compaction is likely to increase and soil test potassium levels are low, it may be necessary to band potassium at planting time for maximum economic yields.

Surface soil pH should be closely monitored when nitrogen fertilizers are surface applied. More frequent applications of lime at low rates may be needed, particularly on sandy soils with low buffering capacities. No adverse effects of surface lime applications have been reported. Very few micronutrient problems have been associated with no-tillage or minimum tillage systems of crop production.

Nitrogen management for no-tillage corn will require the greatest attention if yields are to be optimized. The preferred placement of nitrogen in any conservation tillage system is injection or incorporation beneath the soil surface. Anhydrous ammonia (82 percent nitrogen) and UAN (28 percent nitrogen) solutions are well suited to this practice and can be applied preplant or sidedress after emergence.

Ammonium nitrate (34 percent nitrogen) is the preferred nitrogen source when nitrogen fertilizer is to be surface applied and not incorporated, because of its lower ammonia volatilization potential compared to urea (46 percent nitrogen) or UAN (28 percent nitrogen). Ammonium sulfate (21 percent nitrogen) also has a low ammonia volatilization potential and may be used as a surface broadcast application; however, it has the disadvantage of having the greatest residual acidity. Its use will require greater attention to surface soil pH.

All nitrogen fertilizers which are surface applied to no-tillage or minimum tillage systems are subject to nitrogen immobilization by the surface residues. When urea and UAN solutions are surface applied they are not only subject to nitrogen immobilization by the residues but are also subject to ammonia volatilization losses. The amount of ammonia lost by volatilization is greatly dependent upon the time of application and occurrence of rain after the application. In Michigan, applications made in April or early May have less potential for loss than applications made in late May or June because of the lower temperatures and higher frequencies of rainfall in early spring. Losses of greater than 30 percent nitrogen can be expected when urea is surface applied to the residue of no-tillage soils, if a warm, dry period of no rainfall for 7 days occurs immediately after the application. Farmers should never apply urea to freshly limed no-tillage soils because the high pH at the surface can greatly increase ammonia volatilization.

Ultimately, the nitrogen fertilizer program a farmer chooses will depend on many factors such as cost and availability of the material, the availability of his own equipment or a custom applicator, and weather conditions prior to and after planting. Corn growers who choose to surface apply their nitrogen to soils that have a large amount of residue on the surface may want to increase their nitrogen rate by 10 to 20 percent to compensate for nitrogen immobilization and potential ammonia volatilization losses.

LITERATURE CITED

Anonymous. 1978. Glossary of soil science terms. Soil Science Society of America. Madison, WI. pp. 36.

Bandel, V.A., S. Dziernia, G. Stanford, and J.O. Legg. 1975. N behavior under no-tillage and conventional corn culture. First-year results using unlabeled N fertilizer. Agron. J. 67:782-786.

Bandel, V.A., S. Dziernia, and G. Stanford. 1980. Comparison of N fertilizers for no-tillage corn. Agron. J. 72:337-341.

Belcher, C.R., and J.L. Ragland. 1972. Phosphorus adsorption by sod planted corn (Zea mays L.) from surface-applied phosphorus. Agron. J. 64:754-756.

Blevins, R.L., and D. Cook. 1970. No-tillage: its influence on soil moisture and soil temperature. Kentucky Agr. Exp. Sta. Prog. Rep. 187.

Blevins, R.L., D. Cook, S.H. Phillips, and R.E. Phillips. 1971. Influences of no-tillage on soil moisture. Agron. J. 63:593-596.

Blevins, R.L., G.W. Thomas, and P.L. Cornelius. 1977. Influence of no-tillage and nitrogen fertilization on certain soil properties after 5 years of continuous corn. Agron. J. 69:382-383.

Darlington, W.H. 1983. The effect of tillage method and fertilization on the yield and elemental composition of corn and soybeans. M.S. Thesis. Michigan State University, East Lansing, MI.

Doran, J.W. 1980a. Microbial changes associated with residue management with reduced tillage. Soil Sci. Soc. Am. J. 44:518-524.

Doran, J.W. 1980b. Soil microbial and biochemical changes associated with reduced tillage. Soil Sci. Soc. Am. J. 44:765-771.

Eckert, D.J. 1983. Phosphorus fertilization of no-tillage corn. Ohio State Soil Fertility Research Report, 1982.

Estes, G.O. 1972. Elemental composition of maize grown under no-tillage and conventional tillage. Agron. J. 64:733-735.

Fink, R.J., and D. Wesley. 1974. Corn yield as affected by fertilization and tillage system. Agron. J. 66:70-71.

Fox, R.H., and L.D. Hoffman. 1981. The effect of N fertilizer source on grain yield, N uptake, soil pH, and lime requirements in no-tillage corn. Agron. J. 73:891-895.

Graff, O. 1969. Activity of earthworms under natural organic materials estimated by sample castings. Pediolgia. 9:120-127.

Hargrove, W.L., J.T. Touchton, and R.N. Gallaher. 1982. Influence of tillage practices on the fertility status of an acid soil double-cropped to wheat and soybeans. Agron. J. 74:684-687.

Johnson, J.W. 1983. Response of no-tillage corn to N, P or K. Ohio State Soil Fertility Research Report, 1982.

Ketcheson, J.W. 1980. Effect of tillage on fertilizer requirements for corn on a silt loam soil. Agron J. 72:540-542.

Lal, R. 1979. Influence of 6 years of no-tillage and conventional plowing on fertilizer response of maize (Zea mays L.) on an alfisol in the tropics. Soil Sci. Soc. Am. J. 43:399-403.

Legg, J.O., G. Stanford, and O.L. Bennett. 1979. Utilization of labeled-N fertilizer on silage corn under conventional and no-tillage culture. Agron. J. 71:1009-1115.

Lutz, J.A., Jr., and J.H. Lillard. 1973. Effect of fertility treatments on the growth, chemical composition and yield of no-tillage corn on orchardgrass sod. Agron. J. 65:733-736.

Mengel, D.B., D.W. Nelson, and D.M. Huber. 1982. Placement of nitrogen fertilizers for no-tillage and conventional till corn. Agron. J. 74:515-518.

Moncrief, J.F., and E.E. Schulte. 1981. The effect of tillage and fertilizer source and placement on nutrient availability to corn. Unpublished data.

Moschler, W.W., and G.D. Jones. 1974. Nitrogen for no-tillage corn in Virginia: sources and time of application. Comm. Soil Sci. Plant Anal. 5(6):547-556.

Moschler, W.W., and D.C. Martens. 1975. Nitrogen, phosphorus and potassium requirements in no-tillage and conventionally tilled corn. Soil Sci. Soc. Amer. Proc. 39:886-891.

Moschler, W.W., D.C. Martens, C.I. Rich, and G.M. Shear. 1973. Comparative lime effects on continuous no-tillage and conventionally tilled corn. Agron. J. 65:781-783.

Moschler, W.W., and G.M. Shear. 1975. Residual fertility in soil continuously cropped to corn by conventional tillage and no-tillage methods. Agron. J. 67:45-48.

Moschler, W.W., G.M. Shear, D.C. Martens, G.D. Jones, and R.P. Wilmouth. 1972. Comparative yield and fertilizer efficiency of no-tillage and conventionally tilled corn. Agron. J. 64:229-231.

Phillips, R.E., et al. 1971. Agronomy Research, Univ. Kentucky. Misc. Publ. 394:36-38.

Phillips, R.E., R.L. Blevins, G.W. Thomas, W.W. Frye, and S.H. Phillips. 1980. No-tillage agriculture. Science 208:1108-1113.

Rice, C.W., and M.S. Smith. 1982. Denitrification in no-tillage and plowed soils. Soil Sci. Soc. Am. J. 46: 1168-1173.

Rice, C.W., and M.S. Smith. 1984. Short term immobilization of ^{15}N in no-tillage and plowed soils. Soil Sci. Soc. Am. J. 48:295-297.

Shear, G.M., and W.W. Moschler. 1969. Continuous corn by the no-tillage and conventional tillage methods: a six year comparison. Agron. J. 61:524-526.

Terman, G.L. 1979. Volatilization of nitrogen as ammonia from surface applied fertilizers, organic amendments and crop residues. Adv. Agron. 31:189-223.

Thomas, G.W., R.L. Blevins, R.E. Phillips, and M.A. McMahon. 1973. Effect of a killed sod mulch on nitrate movement and corn yield. Agron. J. 65:736-739.

Touchton, J.T., and W.L. Hargrove. 1982. Nitrogen sources and methods of application for no-tillage corn production. Agron. J. 74:823-826.

Triplett, G.B., Jr., and D.M. Van Doren, Jr. 1969. Nitrogen, phosphorus and potassium fertilization of non-tilled maize. Agron. J. 61:637-639.

CHAPTER 9

THE EFFECTS OF CONSERVATION TILLAGE ON SOIL PROPERTIES

R.L. Blevins, W.W. Frye, and M.S. Smith
Department of Agronomy
University of Kentucky
Lexington, Kentucky 40546-0091

INTRODUCTION

Soil properties change when farmers change from a system of agricultural production that includes moldboard plowing and numerous tillage operations, such as conventional tillage, to a reduced tillage system. The extent of change is determined by the degree of reduction in tillage, amount of residue returned to the soil surface and length of time that the tillage system is practiced. The most extreme differences are observed in comparing no-tillage planting into a sod or cover crop as compared to conventional tillage where a turn-plow is used followed by multiple secondary tillage operations. Other tillage practices such as chisel plowing, plow-plant and disking are intermediate in tillage intensity and intermediate in how they affect soil properties.

The change from intensive to reduced tillage systems can be expected to result in certain changes in the soil environment. Soil properties most commonly affected are soil water retention and movement, soil density and porosity, infiltration, structure, organic matter, pH, microbial activity and nutrient distribution and availability.

The investigation reported in this paper (No. 84-3-40) is a project of the Kentucky Agricultural Experiment Station and is published with approval of the Director.

PHYSICAL PROPERTIES

Conservation tillage practices that leave mulch material on the soil surface have a positive influence on several soil physical properties including increased organic matter, better infiltration, more stable aggregates, and conservation of soil moisture.

Soil Water

Conservation of soil water is one of the major advantages of conservation tillage (Phillips et al., 1980). The mulch at the surface of the soil reduces soil water evaporation resulting in more water stored for plant use (Figure 1). This means more efficient use of soil water for conservation tillage systems in terms of yields as compared to conventional tillage. Maintaining all or part of the previous crop residue at the soil surface shades the soil, serves as a vapor barrier against water losses from the soil, slows surface water runoff and usually increases infiltration (Triplett et al., 1968).

Numerous research studies (Triplett et al., 1968; Jones et al., 1969; Blevins et al., 1971; Lal, 1981) have shown that water is conserved in the upper soil profile in no-tillage systems (Figure 1). This system significantly reduces evaporation losses in the early part of the cropping season before the crop canopy closes. Thus, the more efficient use of available soil moisture in a conservation tillage system increases crop yield potential.

Overall, the generally higher soil water content under conservation tillage is usually beneficial but can occasionally be detrimental. Excessive moisture in poorly drained or slowly permeable soils may promote denitrification losses (Rice and Smith, 1982). Lysimeter studies in Kentucky (Tyler and Thomas, 1977) showed deeper movement of nitrate in no-tillage than conventional tillage early in the corn (Zea mays) growing season. Deeper leaching is enhanced by greater continuity of pores in the less disturbed soil of the no-tillage system which allows more rapid movement of water and nitrate through the macropores. The wetter soil under no-tillage means more pores are filled with water; so when rainfall occurs, a greater proportion of the water moves through macropores and leaches soluble nutrients, such as nitrate, to a greater soil depth.

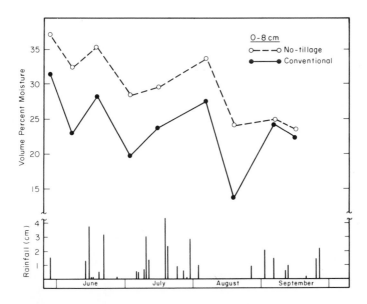

Figure 1. Volume percent soil moisture distribution in the 0 to 8 cm soil depth under no-tillage and conventional tillage (Thomas et al., 1973).

Structure, Density and Porosity

　　　　Soil density and porosity are soil properties that have a profound influence on air and water movement and potential soil productivity. Changes in soil density as affected by conservation tillage vary when compared over a range of agro-climatic zones. Long-term studies in Kentucky and Virginia on medium textured soils show no net change in bulk density (Table 1); however, data from Gantzer and Blake (1978) show a higher bulk density for no-tillage, and this agrees with studies by Cannell and Ellis (1977) on clayey soils in England.

　　　　It seems that the long-term effect of no-tillage on soil parameters such as bulk density and porosity is closely related to soil types and climatic conditions. On medium textured soils such as those studied in Kentucky and Virginia (Table 1), no-tillage has no detrimental effect on bulk density. Wheel-track compaction appears to be effectively ameliorated by the freezing-thawing cycles in the winter months and by increased root proliferation near the soil surface.

Table 1. Soil and bulk density under different tillage systems (Blevins et al., 1983; Gantzer and Blake, 1978; Shear and Moschler, 1969).

Tillage system	Bulk density, g/cm^3		
	Kentucky	Virginia	Minnesota
No-tillage	1.25	1.48	1.25
Conventional tillage	1.29	1.43	1.12[a]

[a]Fall-plowed in 1973; sampled in September, 1974.

Additions of mulches on Alfisols in Nigeria (Lal, 1976) resulted in improved soil porosity, soil structure and water transmission. Triplett et al. (1968), in Ohio, reported an increase in infiltration capacity of soil under no-tillage management; they attributed this to a greater concentration of organic matter at the surface. Research in Germany (Ehlers, 1975) showed better air and water transmission in non-tilled soils. Ehlers suggested that the improved soil environment was due to increased earthworm activity and the maintenance of pore continuity from the surface to lower soil depths.

Under some soil-climate situations, conservation tillage may influence soil physical properties in a negative way. Minnesota workers (Gantzer and Blake, 1978) reported significantly higher bulk density with no-tillage treatments as compared to conventional tillage on a fine-textured soil. Cannell and Ellis (1977), working with direct drilled wheat in Great Britain, found a reduction in total soil pore space as compared to conventionally prepared land. They concluded that soil aeration was improved under no-tillage on some soil types (medium-textured, well-drained) but increased density and decreased aeration was found on clayey soils during wet winters.

Water stable aggregation is another parameter sometimes used to evaluate soil structure. Increased organic matter content often correlates well with improved soil aggregation. The soil aggregation index was measured in Indiana (Mannering et al., 1975) to compare several conservation tillage systems with conventional tillage. The results shown in Table 2 are averages of four different soils. These

Table 2. Effect of tillage on soil aggregation of four Indiana soils after 5 years of cropping under different tillage systems (Mannering et al., 1975).

Tillage System	Aggregation index with soil depth	
	0 to 5 cm	5 to 15 cm
Conventional tillage	0.35	0.47
Chisel plow	0.46	0.56
No-tillage	0.77	0.70

data further substantiate the principle that increased tillage results in decreased soil aggregation.

Organic Matter Content in Soil

Plowing and other tillage operations increase the rate of organic matter decomposition. So, it is not surprising that soils have higher organic matter content after being in conservation tillage for several consecutive years. Also, there is a shift in distribution of organic matter with the build-up occurring mostly in the 0 to 5 cm surface layer (Table 3). Doubling, for example, the amount of

Table 3. Soil organic matter distribution after 5 years (1975) and 10 years (1980) of no-tillage (NT) and conventional tillage (CT) corn production.

Soil depth (cm)	Organic matter content, percentage			
	1975		1980	
	NT	CT	NT	CT
0 to 5	4.11	2.78	4.82	2.40
5 to 15	2.15	2.60	2.34	2.31
15 to 30	1.24	1.47	1.15	1.22

organic matter in the surface of non-tilled soils could result in very different biological, chemical and physical properties. There is also a relationship between organic matter accumulation and added fertilizer nitrogen. A comparison of a limed and unlimed soil under no-tillage that received four nitrogen rates for 10 consecutive years of corn production is shown in Figure 2. Soil organic matter increased with increasing rates of nitrogen fertilizer on both limed and unlimed soil but increased more rapidly and to a higher level in unlimed soil. This was probably a result of decreased microbial activity in the unlimed soil. A combination of no-tillage management and medium to high nitrogen fertilizer appears to be a management system that will maintain the organic matter level and possibly increase it over time.

The increased organic matter content near the soil surface plus the decomposed mulch layer results in a soil zone that has a higher moisture content, usually an increased number of microbes, and increased microbial activity. Fertilizer

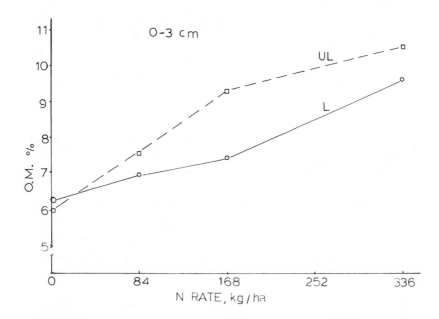

Figure 2.   Influence of nitrogen rates on organic matter accumulation in the 0 to 3 cm depth of limed (L) and unlimed (UL) soil after 10 years of no-tillage corn production.

nitrogen broadcast on the soil surface may be subject to denitrification (Doran, 1980; Rice and Smith, 1982), since a greater number of denitrifying microbes occur near the surface of the non-tilled soil. The nitrogen may also be rapidly taken up by microbes and immobilized (Smith and Rice, 1983; Rice and Smith, 1984). This immobilization may result in nitrogen deficiency in the crop, especially if low rates of nitrogen are used.

Researchers in Indiana, Kentucky and Maryland have reported an increase in nitrogen fertilizer efficiency in the no-tillage system if fertilizer is placed below the soil surface in a band to the side of the row and 8 to 10 cm deep. Delayed application of nitrogen fertilizer may also improve its efficiency in conservation tillage systems (Frye, 1984).

CHEMICAL PROPERTIES

Changing from intensive tillage systems to conservation tillage means less mixing of surface-applied amendments and organic residues. One result is accumulation of certain nutrients near the soil surface if the conservation tillage system is continuous over a period of several years.

Soil Acidity

The most obvious change in soil chemical properties resulting from the small amount of soil disturbance and mixing is the acidification of the soil surface where nitrogen fertilizer was applied. The acidification or reduction in pH takes place at a more rapid rate in no-tillage than conventional tillage. The acidification is closely related to the amount of nitrogen fertilizer applied (Figure 3). For this reason, the rapid lowering of pH associated with reduced tillage is less of a problem with crops such as soybeans (Glycine max) or other crops that demand less nitrogen fertilizer. If the soil pH is allowed to continue declining, a decrease in calcium and magnesium and an increase in aluminum and manganese can be expected (Figure 4). Aluminum toxicity may become a serious threat to crop production. Low pH levels near the soil surface can lead to crop yield losses due to nutrient imbalance and additional weed competition. An acid soil surface enhances rapid deactivation of triazines, a herbicide commonly used for corn production (Kells et al., 1980). The acidification problem can be easily corrected or prevented by timely applications of lime to the soil surface (Blevins et al., 1978).

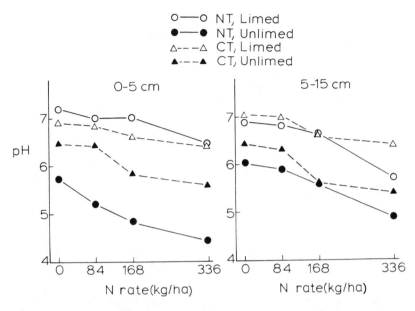

Figure 3. Influence of tillage system and nitrogen rates on changes in soil pH after 10 years of corn production (from Blevins et al., 1983).

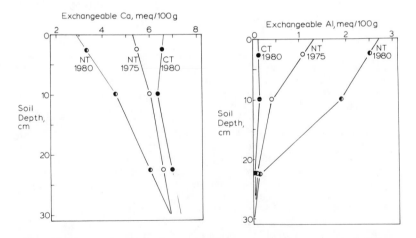

Figure 4. Exchangeable calcium and aluminum distribution with soil depth after 5 years (1975) and 10 years (1980) no-tillage and conventional tillage corn production (Blevins et al., 1977; Blevins et al., 1983).

Phosphorus and Potassium

If conservation tillage is continued for several years and phosphorus and potassium fertilizer are added annually, a build-up of these nutrients will occur in the 0 to 5 cm layer. This distribution of phosphorus and potassium does not appear to be a problem and may increase phosphorus availability. The additional water in the surface zone enhances phosphorus diffusion to the plant roots and encourages root proliferation near the soil-mulch interface.

Organic Nitrogen

Soil organic nitrogen distribution follows a pattern similar to other nutrients and organic matter content in that it accumulates near the soil surface with no-tillage (Table 4). A comparison of soil organic nitrogen in a long-term study in Kentucky (Blevins et al., 1983) showed an increase of organic nitrogen in the 0 to 5 cm surface layer of soil under no-tillage, while under conventional tillage, the soil contained lower levels of organic nitrogen

Table 4. Comparison of soil organic nitrogen after 10 years of no-tillage (NT) and conventional tillage (CT) corn production using four nitrogen rates (from Blevins et al., 1983).

Soil depth (cm)	N rate (kg/ha)	Organic N, percentage	
		NT	CT
0 to 5 cm	0	0.23	0.14
	84	0.25	0.15
	168	0.24	0.15
	336	0.30	0.15
5 to 15	0	0.15	0.14
	84	0.15	0.14
	168	0.15	0.16
	336	0.16	0.16

which was evenly distributed throughout the plow layer (0 to 15 cm). Organic nitrogen increased with increased nitrogen fertilizer rates in the no-tillage. Greater immobilization probably accounted for most of the build-up.

SUMMARY

The type of tillage system used affects both the physical and chemical soil properties. Early in the growing season the upper part of the soil is usually wetter under conservation tillage with a mulch on the soil surface than under conventional tillage where the soil is bare. In most situations, this is desirable because it is beneficial to germination and plant growth; however, it may contribute to critically low soil temperature under certain climatic conditions. Infiltration is usually greater and soil aggregation is improved in reduced tillage systems. Soil compaction is not a problem for most medium textured soils but may be a problem for easily compacted clayey soil. Organic matter accumulates near the soil surface due to lack of mechanical mixing and a slower rate of decomposition under conservation tillage management.

Changes in chemical properties that occur under conservation tillage include higher organic nitrogen, lower pH and exchangeable calcium and magnesium, higher levels of exchangeable aluminum and manganese, lower nitrate concentrations, and higher levels of available phosphorus and potassium. Most of these changes are positive attributes. The negative ones, such as lower pH, are usually easy to correct and manage.

If farmers and researchers have an understanding of how the soil properties in conservation tillage differ from conventionally-tilled soils, management inputs can be modified to improve productivity. This may require more efficient management of nitrogen fertilizers and more frequent soil testing to provide a sound basis for determining lime and fertilizer needs.

LITERATURE CITED

Blevins, R.L., D. Cook, S.H. Phillips, and R.E. Phillips. 1971. Influence of no-tillage on soil moisture. Agron. J. 63:593-596.

Blevins, R.L., G.W. Thomas, and P.L. Cornelius. 1977. Influence of no-tillage and nitrogen fertilizer on certain soil properties after five years of continuous corn. Agron. J. 69:383-386.

Blevins, R.L., L.W. Murdock, and G.W. Thomas. 1978. Effect of lime application on no-tillage and conventionally tilled corn. Agron. J. 70:322-326

Blevins, R.L., G.W. Thomas, M.S. Smith, W.W. Frye, and P.L. Cornelius. 1983. Changes in soil properties after 10 years non-tilled and conventionally tilled corn. Soil and Tillage Res. 3: 135-146.

Cannell, R.Q., and F.B. Ellis. 1977. Review of progress on research on reduced cultivation and direct drilling. Agric. Res. Council Letcombe Lab. Annual Rep. 1976:25-27.

Doran, J.W. 1980. Microbial changes associated with residue management with reduced tillage. Soil Sci. Am. J. 44:518-524.

Ehlers, W. 1975. Observations on earthworm channels and infiltration on tilled and untilled loess soil. Soil Sci. 119:242-249.

Frye, W.W. 1984. Energy requirement in no-tillage. Pages 127-151 in R.E. Phillips and S.H. Phillips, eds. No-tillage Agriculture, Principles and Practices. Van Nostrand Rheinhold, New York, NY. 306 pp.

Gantzer, C.J., and G.R. Blake. 1978. Physical characteristics of LeSueur clay loam soil following no-till and conventional tillage. Agron. J. 70:853-857.

Jones, J.N., Jr., J.E. Moody, and J.H. Lillard. 1969. Effects of tillage, no-tillage, and mulch on soil water and plant growth. Agron. J. 61:719-721.

Kells, J.J., C.E. Rieck, R.L. Blevins, and W.M. Muir. 1980. Atrazine dissipation as affected by surface pH and tillage. Weed Sci. 28:101-104.

Lal, R. 1976. No-tillage effects on soil properties under different crops in western Nigeria. Soil Sci. Amer. Proc. 40:762-768.

Lal, R. 1981. No-tillage farming in the tropics. Pages 103-151 in R.E. Phillips, G.W. Thomas, and R.L. Blevins, eds. No-Tillage Research: Research Reports and Reviews. Univ. of Kentucky, College of Agriculture and Agricultural Experiment Station, Lexington, KY. 151 pp.

Mannering, J.V., D.R. Griffith, and C.B. Richey. 1975. Tillage for moisture conservation. Paper No. 75-2523. Amer. Soc. Agric. Eng., St. Joseph, MI.

Phillips, R.E., R.L. Blevins, G.W. Thomas, W.W. Frye, and S.H. Phillips. 1980. No-tillage agriculture. Science 208:1108-1113.

Rice, C.W., and M.S. Smith. 1982. Denitrification in plowed and no-till soils. Soil Sci. Soc. Am. J. 46:1168-1173.

Rice, C.W., and M.S. Smith. 1984. Short-term immobilization of fertilizer N at the surface of no-till and plowed soils. Soil Sci. Soc. Am. J. 48:295-297.

Shear, G.M., and W.W. Moschler. 1969. Continuous corn by no-tillage and conventional tillage methods: a six-year comparison. Agron. J. 61:524-526.

Smith, M.S., and C.W. Rice. 1983. Soil biology and biochemical N transformations in no-till soils. Pages 215-226 in W. Lockeretz, ed. Environmentally Sound Agriculture. Praeger Publishers, New York, NY. 426 pp.

Thomas, G.W., R.L. Blevins, R.E. Phillips, and M.A. McMahon. 1973. Effect of a killed sod mulch on nitrate movement and corn yield. Agron. J. 65:736-739.

Triplett, G.B., Jr., D.M. Van Doren, Jr., and B.L. Schmidt. 1968. Effect of corn (Zea mays L.) Stover mulch on no-tillage corn yield and water infiltration. Agron. J. 60:236-239.

Tyler, D.D., and G.W. Thomas. 1977. Lysimeter measurements of nitrate and chloride losses from soil water under conventional and no-tillage corn. J. Environ. Qual. 6:63-66.

CHAPTER 10

A COMPARISON OF CONSERVATION TILLAGE SYSTEMS
FOR REDUCING SOIL EROSION[1]

W.C. Moldenhauer
USDA-ARS, National Soil Erosion Laboratory
West Lafayette, Indiana 47907

INTRODUCTION

There has been a great upsurge recently in the use of non-moldboard plow systems for primary tillage. In 1981, conservation tillage was used on one-third of the harvested cropland in the Corn Belt. The percentage of cropland in no-till was near the national figure of 2.5 percent. The remainder was mainly chisel-plowed or disked with a heavy, primary tillage disk (Crosson and Brubaker, 1982). These systems have become popular because they help reduce erosion, take less time, require less energy, and are generally easier to handle than the moldboard plow-disk-harrow system. Many people have been led to believe that because these soil surfaces are not inverted that they are less erodible. This is the case only when surface residue has been left from the previous crop or from other sources. The following review shows the relationship between erosion reduction and the amount of residue left on the surface from the previous crop in a number of tillage systems.

───────────────

[1]Contribution from USDA Agricultural Research Service, in cooperation with Purdue University Agricultural Experiment Station.

SURFACE RESIDUE LEFT BY VARIOUS TILLAGE METHODS

The effectiveness of any tillage method for controlling erosion ultimately depends upon the amount of crop residue left on the soil surface (Laflen et al., 1980). With a chisel plow, for example, the amount of residue left depends upon the type of chisel used and the type of crop residue. Following corn, the residue left may vary from 10 to 20 percent with 10 cm (4-inch) twisted shanks to 50 percent or more with narrow points. Much less surface residue is left after chiselling soybean land (Table 1).

Regardless of how much residue is left after primary tillage, what is more important in the Corn Belt is the amount remaining after planting. The first 60 days after planting normally is the period of the most rain, as well as the most intense rainstorms.

The percentage of cover left on chiselled ground after planting depends upon how many secondary tillage operations are carried out in the process of smoothing the field surface and incorporating herbicide. Because tillage with 10 cm twisted shanks leaves little residue, there is little difference in erosion control after planting between this method and the moldboard plow, which leaves less than 10 percent of the residue from cornstalks on the surface. Any chiselling on the contour is much more effective in controlling erosion than chiselling up-and-down slope since chisels tend to leave a series of ridges and furrows.

Following primary tillage with straight chisel points or with sweeps, each secondary tillage operation with

Table 1. Crop and tillage effects on surface cover and soil loss, Morly clay loam, 4 percent slope (Mannering and Fenster, 1977).

| Tillage | Percent Cover | | Soil Loss After | |
| | Soybeans | Corn | Soybeans | Corn |
			(metric tons/ha)	
No-till	26	69	13.	2.4
Chisel (up-and-down)	12	25	30.3	15.0
Plow	1	7	40.0	21.8

a disk in the spring cuts the percentage of cover in half (Sloneker and Moldenhauer, 1977). With 50 percent cover left in the fall, therefore, a spring disking would leave about 25 percent and a subsequent disking to incorporate herbicide would reduce the residue cover to 10 or 12 percent.

EROSION CONTROL BY VARIOUS PERCENTAGES OF RESIDUE COVER

Even small percentages of cover reduce erosion. For example, Laflen et al., (1980), show that for each 10 percent increase in groundcover, erosion is reduced about 40 percent. The greatest reduction in erosion comes between 0 and 20 percent cover. A 65 percent reduction in soil loss was achieved with 20 percent surface cover. Johnson and Moldenhauer (1979) show similar to even greater reductions from small percentages of residue cover.

More conservatively, based on results from a number of experiments, Wischmeier and Smith (1978) reported a 20 percent reduction in erosion for each 10 percent increase in groundcover. A 36 percent reduction in erosion was achieved with 20 percent cover. With 30 percent cover, the reduction was 48 percent.

EFFECT OF TILLAGE SYSTEMS ON EROSION

One problem with straight-row farming is the up-and-down-hill tillage marks, especially from chisels or ammonia applicator knives. Much chisel plowing in the Corn Belt has little effect on reducing soil erosion because of the small amount of residue left after planting and the up-and-down-hill chisel marks. This is particularly serious in chiselled soybean stubble as shown in Table 1. Some manufacturers of chisel plows are aware of the effect up-and-down-hill chisel marks have on runoff and erosion. One has attached various smoothing devices behind the chisel plow to erase these marks and leave the residue better distributed. This also has the effect of smoothing the chiselled ground and reducing or eliminating the need for secondary tillage before planting with a fluted coulter. On the other hand, smoothing reduces the roughness and surface porosity effect, which helps reduce runoff and erosion. Research and development are needed to make chisel plow tillage a more effective conservation tillage system.

Ridge-till, strip-till, and disk-tillage all effectively reduce soil losses, depending upon how much residue remains on the soil surface. With all of these systems, however, it is important to avoid planting up-and-down hill if possible. Their erosion control effectiveness drops greatly when tillage operations are not conducted across the slope or on the contour. Ridge-tilling, when done correctly, allows runoff to move down the ridges and through the residue accumulated in the furrow bottoms. The result is very effective soil erosion control following a corn crop. The reduced amount of residue from soybeans and its quicker decay offer less protection following soybeans.

The correlation between percentage of residue cover and soil loss applies equally to no-till, ridge-till, strip-till, disk-till and all other systems. No-till is generally the most effective means of erosion control, mainly because more residue is left. The longer a field remains in no-till, the more effective erosion control becomes (Wischmeier, 1973, and Van Doren et al., 1984). This better erosion control is related to improved soil structure and better infiltration. Also, detachment by raindrops becomes more difficult because soil aggregates become larger and more stable. With no-till on moderate slopes (2-5%), row direction affects soil loss very little. Up-and-down-hill planting is nearly as effective as contouring. On slopes greater than 10 percent, the coulter slot can erode, especially after a soybean crop.

Tests at the Milan Experiment Station in West Tennessee (Shelton et al., 1983) show that soil losses are a function of vegetative cover (Table 2). The effect of no-till

Table 2. Effects of soybean tillage and cropping systems on soil loss from a 63 mm rainfall at Milan, Tennessee, on June 11, 1981.[2] (Moldenhauer et al., 1983).

Tillage and Cropping System	Soil Loss (metric tons/ha)
Conventional tillage, single-cropped soybeans, no winter coat	62
Conventional tillage, double-cropped wheat-soybeans	0.3
Disk tillage, single-cropped soybeans, no winter cover	84
No-tillage, single-cropped soybeans	0.8
No-tillage, double-cropped wheat-soybeans	0.3

[2]Source: Personal communication with Shelton, Tompkins, and Tyler.

Table 3. Effect of tillage on runoff and soil loss resulting from a 31 mm rainfall on corn watersheds in Clark County, Kentucky, November 25, 1982[2] (Moldenhauer et al., 1983).

Tillage System[3]	Runoff (mm)	Soil Loss (metric tons/ha)
Conventional tillage	6.0	2.3
Chisel plow and disk	2.7	0.3
Disk	0.1	trace
No-tillage	0	0

in controlling soil erosion is also apparent when comparing soil loss for single-cropped soybeans using conventional tillage (62 metric tons/ha) and single-cropped soybeans under no-till (0.8 metric tons/ha).

In an erosion study underway in Clark County, Kentucky, runoff and soil loss are being compared for corn grown under no-till, conventional tillage, chisel plowing followed by disking, and disking alone. The effectiveness of no-till in controlling soil loss was documented during a rainfall of 31 mm on November 25, 1982 (Table 3). Residue left on the soil surface with no-till and conventional tillage was 4.9 and 0.7 metric tons/ha, and soil loss was 0 and 2.3 metric tons/ha, respectively.

Laflen et al., (1978), measured erosion from several conservation tillage systems and compared it with the conventional (spring moldboard plow-double disk) system (Table 4). Actual soil losses for the conventional system were 40, 50 and 62 metric tons/ha for Tama, Kenyon, and Ida soils, respectively. Only no-till coulter planting effectively reduced soil erosion on the Ida soil, and it was the only system that kept soil losses close to acceptable limits on the other two soils. Disking was more effective than chisel plowing on Tama and Ida soils.

---

[2]Source: Personal communication with Aswad, Bitzer, and Blevins.

[3]Watershed slope, 12 percent.

Table 4. Percent of soil loss from several conservation tillage systems compared with conventional systems[5] (Laflen et al., 1978).

Soil	No-till	Disk (% conventional)	Chisel
Tama silty clay loan (Typic Argiudolls)	8	25	48
Kenyon loam (Typic Hapludolls)	15	37	37
Ida silt loam (Typic Udorthents)	10	46	80

Wischmeier (1973) reported an experiment comparing soil loss from several conservation tillage systems with the conventional moldboard plow and disk system (Table 5). The most effective system in reducing soil loss was the fourth year of no-till coulter planting on the contour which had a soil loss only 5 percent that of conventional. Equally as effective was chisel plowing with sweeps. Chisel plowing with points (up-and-down slope) lost 0.52 as much soil as conventional in crop stage No. 1 (planting to 30 days after) and by crop stage No. 2 (30-60 days after planting) had lost 0.71 as much as conventional tillage. Soil loss from no-till planting did not show this increase from crop stages No. 1 and No. 2. Wischmeier (1973) compared the effectiveness of several tillage conditions in reducing erosion. He also discussed the failure of mulch because of floating off or being undercut by flowing water when critical slope-length limits are exceeded.

Van Doren et al., (1984), reported an experiment comparing soil loss from three tillage systems after 18 years of continuous corn (Table 6). Since the experiment was done in August, there was no difference between the plow-disk-cultivate and the plow-cultivate (popularly known as plow-plant) tillages. Soil loss from these two tillages was 18 fold greater than from no-tillage where all residues were left on the soil surface.

---

[5]Average slope was 4.7, 4.8, and 12.2 percent for the Tama, Kenyon and Ida soils, respectively; rows were up-and-down slope, and testing was done 11 to 35 days after planting; the crop was corn following corn.

Table 5. Soil loss as percentage of corresponding loss from conventional tillage (Wischmeier, 1973).

Practice	Row Direction	Crop Stage[5]	Average Soil Loss (% of Conventional)
No till			
1st year	contour	1	15
1st year	up and down	1	18
1st year	up and down	2	14
2nd year	contour	1	11
2nd year	contour	1	5
Disk	contour	1	19
Disk	up and down	1	35
Disk	up and down	2	37
Sweep till	contour	1	6
Chisel	up and down	1	52
Chisel	up and down	2	71

Table 6. Soil loss from various tillages after 18 years of continuous corn--Wooster, Ohio (Van Doren, et al, 1984).

Tillage	Cover (%)	Soil Loss (metric tons/ha)
No-till	56	2.6a
Plow-disk-cultivate	5	46.5b
Plow and cultivate	5	45.0b

a. Average of two plots
b. Average of three plots

[5]Crop stage 1 was the period from planting to 30 days after. Crop stage 2 was from 30 to 60 days after planting.

EFFECT OF SOYBEANS ON EROSION

Soybeans in a corn-soybean rotation greatly reduce the effectiveness of no-till in controlling erosion (Table 1). Residue remaining after soybeans is much less than after corn or small grain, and soybean residue breaks down more rapidly in comparison with corn or wheat residue. For example, on slopes 300 feet long, a corn-soybean rotation effectively controls erosion on soils of moderate erodibility with slopes up to 4 percent in the western Corn Belt. With continuous corn, no-tillage will protect soils with slopes up to 6 percent under similar conditions.

Cover crops, such as legumes, certain grasses, and wheat or rye seeded into the senescing soybean crop, show promise for helping to control erosion following soybeans. The cover crop is killed with a contact herbicide in the spring and planting is done into the soybean and cover crop residue.

STATUS OF CONSERVATION TILLAGE IN THE CORN BELT

What is the status of conservation tillage in the Corn Belt? Chiselled corn residue is much more common now than a few years ago. Chiselled bean residue is widely accepted. Use of no-till, or a shallow pass with a field cultivator, for corn following soybeans is gaining acceptance rapidly. There is great interest in till-planting on ridges. It remains to be seen how many will actually accept this practice.

Much of the interest in conservation tillage is the result of an economic situation that demands a reduction in costs. Research and farmer experience have shown the systems to be workable. Better equipment and chemicals are available, and equipment is often available on loan through Conservation Districts, the Extension Service in some states, and industry. There also is greater awareness of erosion problems and the fact that conservation tillage can solve them. Although the shift away from moldboard plow tillage systems is encouraging, there is a need to leave more crop residue at the surface, especially when using chisel plow systems.

SUMMARY

There has been a tremendous upsurge in the use of non-moldboard plow tillage systems in the last fifteen years. One reason for this is that surface residue is very effective in reducing erosion. Another is that many times these systems are easier to manage, require less time and are less expensive than the moldboard plow-disk-harrow system. It must be emphasized that surface residue is the key to reducing erosion with these non-moldboard plow systems. While surface roughness, porosity, and other things are important, it is the surface residue that maintains the high infiltration rate by preventing surface sealing. Without surface residue, chiselling, disking and other systems are little better than moldboard plowing.

While no-till is the most effective erosion control system because of the generally higher rates of surface residue, the formerly plowed layer, under certain soil-climatic conditions, may become fairly dense, causing high runoff. Also, crops such as soybeans, cotton, potatoes, and sugar beets leave little residue. With little or no residue, a no-till system is actually poorer than a tilled system for erosion control. In such a situation the tilled system is preferable because of the high porosity in the early states after tillage. In some soils, a combination of surface residue and high porosity is required for most effective erosion reduction. In these cases some type of tillage is required. Most effective in this situation would be sweeps, a deep subsoiler or a Paraplow--some implement that would leave most of the residue on the surface.

LITERATURE CITED

Crosson, P. and S. Brubaker. 1982. Resource and environmental effects of U.S. agriculture. Resources for the Future, Washington, D.C. 20036.

Johnson, C.B. and W.C. Moldenhauer. 1979. Effect of chisel vs moldboard plowing on soil erosion by water. Soil Sci. Soc. Amer. Jour. 43:177-179.

Laflen, J.M., J.L. Baker, R.O. Harwig, W.F. Buchele, and H.P. Johnson. 1978. Soil and water loss from conservation tillage systems. Trans. Am. Soc. Agric. Eng. 21:881-885.

Laflen, J.M., W.C. Moldenhauer, and T.S. Colvin. 1980. Conservation tillage and soil erosion on continuously rowcropped land. In: Crop Production with Conservation in the 80s. Am. Soc. Agr. Eng., St. Joseph, Michigan.

Mannering, J.V. and C.R. Fenster. 1977. Vegetative water erosion control for agricultural areas. In: Soil Erosion and Sedimentation. Am. Soc. Agr. Eng., St. Joseph, Michigan.

Moldenhauer, W.C., G.W. Langdale, W. Frye, D.K. McCool, R.I. Papendick, D.E. Smika, and D.W. Fryrear. 1983. Conservation tillage for erosion control. J. Soil Water Cons. 38:144-151.

Shelton, C.H., F.D. Tompkins, and D.D. Tyler. 1983. Soil erosion from natural and simulated rainfall on five soybean tillage systems. J. Soil Water Cons. 38:425-428.

Sloneker, L.L. and W.C. Moldenhauer. 1977. Measuring the amounts of crop residue remaining after tillage. J. Soil Water Cons. 32:231-236.

Van Doren, D.M., Jr., W.C. Moldenhauer, and G.B. Triplett, Jr. 1984. Tillage systems for optimizing crop production and maintaining soil productivity. Soil Sci. Soc. Amer. Jour. 48:636-640.

Wischmeier, W.H. 1973. Conservation tillage to control water erosion. p. 133-141. In: Conservation Tillage: The Proceedings of a National Conference. Soil Conserv. Soc. of Am., Ankeny, Iowa.

Wischmeier, W.H. and D.D. Smith. 1978. Predicting rainfall erosion losses: A guide to conservation planning. Agr. Handbk. No. 537. Sci. and Educ. Admin., U.S. Dept. Agr., Washington, D.C. 58 pp.

PART II

CONTROL OF WEEDS, INSECTS, AND PLANT DISEASES

CHAPTER 11

CONSERVATION TILLAGE AND WEED CONTROL

J.J. Kells and W.F. Meggitt
Department of Crop and Soil Sciences
Michigan State University
East Lansing, Michigan 48824

Weed control is often considered one of the major problems associated with conservation tillage systems. Adequate weed control is essential to successful crop production. As tillage is decreased, the reliance on chemical weed control becomes greater. In addition, several soil characteristics change in response to tillage systems. Approaches to weed control must be modified when tillage systems are changed.

Conservation tillage involves tillage practices that leave part or all of the previous crop residue on the soil surface to reduce the potential for water and wind erosion. Conservation tillage practices often include the use of a chisel plow or other primary tillage tools, other than a moldboard plow. No-tillage is a form of conservation tillage.

Tillage has a significant impact on weeds. Seed germination of most annual weed species is favored by tillage. Weed seeds have several requirements for germination including moisture, oxygen, proper temperature, and light. By disturbing the soil, weed seeds are often stimulated to germinate by being positioned in a more favorable environment to break dormancy. This is generally observed as a flush of germination following tillage. As tillage is reduced, total weed germination often decreases. In no-tillage, weed germination will be more sporadic and extended over a longer period of time. Rather than a flush of weeds following tillage, weed germination will follow environmental factors such as rainfall.

Along with changes in total weed population, the spectrum of weeds present is likely to shift as tillage practices are changed. Several studies have reported an increase in certain perennial weeds under continuous no-tillage. These include such species as hemp dogbane, common milkweed, yellow nutsedge, and quackgrass, along with many other perennials. One logical approach to deal with an increase in perennials would be to rotate the field away from no-tillage for one year. This may be very effective for controlling simple perennials such as dandelions. However, a recent study at Iowa State University showed that moldboard plowing for one year did not reduce the population of hemp dogbane. Intensive tillage was required for three years before a reduction in population was observed. Perennial weeds may mature more uniformly under no-tillage. This may allow for more complete chemical control in certain cases.

In addition to an increase in perennial weeds, annual grasses have been found to increase as tillage is reduced while annual broadleaved weeds tend to decrease. Weed pressure and spectrum is likely to shift and should be monitored closely under conservation tillage systems.

Another important consideration is the influence of conservation tillage on herbicide effectiveness. Under conservation tillage systems, some or all of the previous crop residue will remain on the soil surface. This surface residue can intercept a herbicide spray and influence its effectiveness. Research at Iowa State University has shown that surface residue does not strongly bind herbicides and most herbicides are quickly washed off the residue by rainfall. However, previous crop residue, especially corn stalks, does influence herbicide effectiveness by inhibiting uniform distribution of the herbicide on the soil surface. A number of steps can be taken to help reduce the potential loss of weed control with high crop residue levels. Increasing the total number of spray droplets per unit area will aid in more uniform spray distribution. This may be accomplished with an increase in spray volume (l/ha). Increasing spray pressure may help to drive the spray through the crop residue and will also produce a larger number of spray droplets per unit area. However, the smaller droplets produced by higher pressure will increase the potential for spray drift. The use of granular herbicide formulations or herbicide impregnated fertilizer may also help to uniformly distribute certain herbicides, since the granules will sift through the residue to the soil surface.

Herbicide rates should also be modified under conservation tillage systems. In general, labeled rates of herbicides should provide adequate weed control. However, since the soil surface is more variable and surface residue is present, herbicides will be more effective if applied at a rate near the higher end of the rate range allowed by the herbicide label.

Herbicide incorporation practices may also need to be modified with conservation tillage systems. With conventional tillage systems, improvements in incorporation equipment have made one-pass incorporation possible. However, when a chisel plow is used for primary tillage, ridges may remain in the field which can inhibit uniform distribution of the herbicide. It is best to level 10 cm or higher ridges prior to application and incorporation of the herbicide. This can be accomplished with a separate pass across the field or with combination tools where the herbicide is applied behind the first tillage tool.

Surface residue can also influence incorporated herbicide effectiveness, primarily by inhibiting uniform incorporation. Most herbicides are not tightly bound to surface residue. However, if the herbicide is applied to a surface with high levels of residue, the herbicide on the residue will not be incorporated uniformly, especially with shallow incorporation. This can be a problem with corn stalks, but is generally not a problem with soybean stubble. In cases where high levels of surface residue are present, a two-pass incorporation program or a combination tool may be needed to incorporate some of the residue prior to herbicide application.

Weed control under no-tillage involves many of the same considerations as previously discussed. Chemical weed control programs involve the same herbicides and herbicide combinations as with other tillage programs, with certain exceptions. Under no-tillage, herbicide options are limited to surface-applied soil treatments and foliar treatments, since incorporation is not possible. In addition, a burndown herbicide is usually needed to remove any vegetation in the field prior to or at planting. As with other conservation tillage practices, it is best to apply herbicides at a rate near the higher end of the rate range. It is also important to use a herbicide program that will provide season long control because weed germination is likely to extend longer into the growing season.

Several burndown materials can be used for vegetation removal in no-tillage. These materials include paraquat, glyphosate, 28 percent liquid nitrogen, and certain soil applied herbicides including cyanazine, linuron or metribuzin. Soil applied herbicides and 28 percent nitrogen will be effective on very small annual weeds only and will show their activity slowly. Results can be expected to be somewhat less consistent. Successful use of these materials will require more intensive management and a complete understanding of the activity of the material being used. Paraquat and glyphosate are the most commonly used vegetation control herbicides in no-tillage. Glyphosate is a systemic herbicide since it will translocate within plants. Paraquat is a contact herbicide and only kills tissue with which it comes in direct contact. Depending on the situation, one of these herbicides may be preferred over the other. For annual weeds or a rye cover crop that has fully tillered, paraquat would be preferred. Glyphosate is the best choice for perennial weeds, rye cover crops not fully tillered, or in a situation of dense foliage.

Several limitations apply to the method of application with these herbicides. When paraquat is used, a fairly high spray volume (187 to 560 l/ha) and pressure (280 kPa) should be used to insure adequate coverage, especially in dense foliage. In addition, this herbicide should not be applied in combination with clay-based fertilizers or phosphate fertilizers. A non-ionic surfactant should be added to the spray tank. Other limitations are outlined on the herbicide label. Glyphosate is best applied in clean water, however, tank mixes with certain herbicides are acceptable. In terms of spray volume, the trend with glyphosate has been to reduce spray volume to enhance herbicidal activity. Glyphosate, a systemic herbicide, does not require complete coverage to kill the plant. However, in dense foliage a higher pressure may be desirable to provide better penetration of the weed canopy in order to insure exposure of all plants present. Glyphosate should not be applied with liquid fertilizer or with dirty water. Paraquat and glyphosate should be selected based on how they fit into the entire crop production system. In certain cases, the best approach to vegetation management in dense foliage may be a treatment with glyphosate one week or more prior to planting, followed by a treatment with paraquat at planting time to remove any small weeds that have recently germinated. Paraquat should never be tank mixed with glyphosate.

The ester formulation of 2,4-D is very effective for killing an alfalfa or clover sod. It should be applied seven to ten days prior to planting no-till corn. This approach is not currently labeled (approved) for soybean production.

Atrazine is a popular option for management of a quackgrass sod prior to and following planting for no-till corn production. The best approach to quackgrass control with atrazine is a sequential treatment of 2.24 kg ai/ha, two weeks or more prior to planting, followed by an additional 2.24 kg ai/ha postemergence. Crop oil concentrate should be included with both treatments. The limitation of this approach to quackgrass management is the rotational restriction associated with atrazine residues.

Surface soil pH should be monitored closely under no-tillage. Where certain forms of nitrogen are surface applied on fields that are not tilled, a rapid and dramatic reduction in surface pH may occur. These include urea, ammonium nitrate, ammonium sulfate, and 28 percent liquid nitrogen. This reduction in surface pH can reduce the effectiveness of certain herbicides, especially triazine herbicides such as atrazine, simazine, cyanazine and metribuzin. The loss of herbicidal activity with very low surface soil pH has commonly resulted in poor weed control. This becomes very dramatic when the pH drops below 5.5. To avoid this problem, the pH of the surface 5 cm of soil should be monitored closely. Low surface soil pH, although it can drastically reduce weed control, can be easily corrected by the addition of lime when needed.

Effective weed control is essential to continuous no-till crop production. Where weed control is effective and no weed seed is produced, the soil reserve of weed seed in the germination zone will be depleted more rapidly than under conventional tillage. Under no-tillage, seeds buried at a depth where seed dormancy will not be broken are not positioned in the germination zone by tillage. In addition, seeds buried below the germination zone by tillage are likely to survive longer since the seeds will be less affected by environmental conditions such as temperature changes, etc. In contrast, where weed seed is produced and the soil is not tilled, all of the seed will remain at or near the soil surface in a position more favorable to break dormancy and stimulate germination. This explains a common observation of no-tillage crop producers that good weed control becomes easier while poor weed control gets worse.

In conclusion, effective weed control in conservation tillage systems is very possible. As tillage systems change, weed problems will change and approaches to weed control must be modified to deal with these changes. The presence of increasing amounts of surface residue may influence herbicide performance, especially under unfavorable environmental conditions. As tillage is reduced, more reliance must be placed on chemical weed control. Since foliar herbicide applications are essentially unaffected by surface residue, postemergence treatments become a more important component of a weed control program as tillage is reduced. Finally, regardless of tillage systems, an important part of a successful weed control program is good crop management. Crop production practices that maximize crop yields such as planting date, population, fertility, etc., will allow the crop to compete more effectively with weeds. Effective weed control involves not only good weed control practices but optimum crop production practices as well.

SUMMARY

Weed control systems require modifications as tillage practices are changed. As tillage is reduced, weed pressure tends to shift with perennial weeds and annual grasses increasing, and annual broadleaved weeds generally decreasing in number. High levels of previous crop residue on the soil surface may influence herbicide effectiveness primarily by inhibiting uniform coverage of the soil surface. Herbicide incorporation can also be influenced by high levels of surface residue. Weed control strategies should be developed to provide consistent control in high residue situations. Under no-tillage, herbicide options are limited to surface soil applications and postemergence treatments. The addition of a burndown herbicide is often necessary to remove vegetation present at planting time. Weed control should be maintained each year to avoid weed seed production and surface pH should be monitored closely. As tillage is reduced, more intensive weed control practices will be needed.

LITERATURE CITED

Cook, W.J. and L.S. Robertson. 1979. Conservation tillage. Cooperative Extension Service, Michigan State Univ. Bull. E-1354, 8pp.

Fawcett, R.S. 1982. Can you control weeds with reduced tillage? In Iowa's 34th Annual Fertilizer and Ag. Chemical Dealers Conf. 12-13 Jan., 1982. Cooperative Extension Service, Iowa State Univ.

Fawcett, R.S. 1983. Weed control in conservation tillage systems. Proc. North Central Weed Control Conf. 38:67.

Griffith, D.R., J.V. Mannering, D.B. Mengel, S.D. Parsons, T.T. Bauman, D.H. Scott, C.R. Edwards, F.T. Turpin, and D.H. Doster. 1982. A guide to no-till planting after

corn or soybeans.   ID-154, Coop. Ext. Ser., Purdue Univ.,
W. Lafayette, IN, 11 pp.

Kells, J.J., C.E. Rieck, R.L. Blevins, and W.M. Muir.
1980.   Atrazine dissipation as affected by surface pH
and tillage. Weed Sci. 28:101-104.

Nelson, J. 1981.   Weed control in conservation tillage.
In Iowa's 33rd Annual Fertilizer and Ag. Chemical Dealers
Conf.   13-14 Jan., 1981.   Cooperative Extension Service,
Iowa State Univ.

Witt, W.W. 1984.   Response of weeds and herbicides under
no-till conditions.   p. 152-170. In NO-TILLAGE AGRICULTURE:
Principles and Practices.   R.E. Phillips and S.H. Phillips. eds.
VanNostrand Reinhold Publishing Co., Inc.

CHAPTER 12

CONSERVATION TILLAGE AND PLANT DISEASE

H. Walker Kirby
Department of Plant Pathology
University of Illinois
Urbana, Illinois 61801

INTRODUCTION

According to Sumner et al. (1981) and Boosalis et al. (1981), conservation tillage and minimum tillage practices may increase, decrease, or have no effect on plant diseases. This is because crop residues may potentially increase the overwintering survival rate of pathogen propagules, may increase the activity of organisms antagonistic to pathogens, may modify the soil environment to favor selected organisms, may affect physical and/or chemical properties of the soil, may influence root growth and activity, may modify soil moisture levels and temperatures, and may affect survival of pathogen vectors. Any or all of these practices will have a direct influence on the level of plant diseases appearing during the next cropping season.

Tillage practices may also indirectly influence plant diseases by affecting the rate and/or uptake of nutrients, the application or types of fertilizers and pesticides, dates of planting, germination rate of seeds and emergence of seedlings, and root distribution. Certain pathogens which depend on crop residues for nutrients and overwintering sites may survive in any conservation tillage system while others may require specialized conditions found only in selected tillage systems combined with certain cropping sequences.

Plant Residue as Survival Habitats for Pathogens

Plant residues from conservation tillage operations provide an excellent source of overwintering inoculum for many field crops. Research has shown that residues provide a means of survival for common plant diseases including several important leaf blights of corn. Boosalis et al. (1967) reported that viable spores of the fungus causing northern corn leaf blight overwintered in corn leaf sheaths, leaf midribs, and on ear husks. However, Burns and Shurtleff (1973) reported in Illinois that survival of the pathogen causing southern corn leaf blight was dependent not only upon the presence of residues but also depended on the race of the fungus. Two races of this leaf blight fungus appeared in Illinois in the early 1970s. Race T was far more severe and attacked both leaf tissue and developing ears; Race O was only able to colonize leaf tissue and was much less damaging. Burns and Shurtleff determined that chiseling or zero-till operations favored survival of Race O when compared to clean plowing. Also, they found that Race T appeared first in the zero-till plots and last in plowed plots. This may be due in part to the ability of the different races to colonize plant residues and to survive the winter in a saprophytic state.

Several bacterial diseases have been found to survive in crop residues (Schuster et al., 1972; Weihling and Vidiver, 1967). Goss's wilt and Holcus leaf spot of corn are two examples where residues directly influence disease outbreaks. Bacteria are generally non-motile and depend on splashing water and/or wind for dispersal rather than actively moving to a host. If infected residues are present in a field during windy, wet weather, bacteria may be splashed onto emerging plants and epidemics may develop.

Soil Factors Affecting Growth and Survival of Pathogens

Crop residues may provide both beneficial and toxic substances to fungi and other pathogens associated with the residues. These substances may provide materials to allow pathogens to exist in a saprophytic state, may inhibit or stimulate germination of overwintering structures and may provide materials for other microorganisms that compete with the pathogens.

A number of studies have documented the effects of diffusing materials from decomposing plant residues and their effect on fungi and other pathogens. Guenzi et al. (1967) observed that wheat and oat residues contained no phytotoxic substances after eight weeks of exposure to field conditions while corn and sorghum required 22 to 28 weeks of exposure before becoming non-toxic. They also determined that phytotoxic products varied among nine wheat varieties. Snyder et al. (1959) determined that the addition of certain crop residues would reduce the activity of a bean root rot fungus. This work indicated that adding wheat straw, corn stover, barley straw, or pine shavings would reduce the level of root rot. However, adding green materials would increase root rot. Their conclusions were that availability of nitrogen to the pathogen affects survival. If residues contain materials that have high carbon to nitrogen ratios, then nitrogen becomes limiting as it is tied up by various organisms and is unavailable to pathogens which may be poor competitors. If materials have low carbon to nitrogen ratios, then pathogen activity may be increased. This has been shown to be particularly true in the case of fungi belonging to the genus Fusarium which are known to be poor competitors that cannot infect without an adequate supply of nitrogen.

Effects on Root Growth and Development

Stress has been shown to be a major contributing factor in the development of many plant diseases. One major component of stress is inadequate root growth or development. The use of conservation tillage systems has been shown to play a major role in the development and distribution of roots in soil.

Chaudhary and Prihar (1974) demonstrated that straw mulch and cultivation enhanced root growth and development in the upper soil layers and also increased the lateral spread of roots. Root development was less in the mulched plots and a more symmetrical root distribution was noted in the cultivated plots. Plots with interrow compaction produced roots that grew deeper but did not spread laterally when compared to other treatments.

Although this study demonstrated that root development was less in mulched plots, plant growth and yields were higher. Several possibilities are noted for this response including lower evaporation rates, reduced soil temperatures and better water storage capacity. All of these factors will reduce stress on the plants and decrease the possibility of crop losses, especially from pathogens that normally parasitize only weakened plants.

Control

Control of plant diseases in conservation tillage systems requires an integrated approach emphasizing alternatives to clean plowing. Although there are many plant pathogens that overwinter in or on residues, the use of resistant varieties and crop rotation offer economical means of reducing potential losses. Leaf blights of corn, several bacteria and other diseases associated with residues are currently controlled in this manner.

One highly successful example of controlling plant diseases with conservation tillage is the "ecofallow" system developed in Nebraska. This tillage technique, discussed by Boosalis et al. (1981) is designed to control weeds and conserve soil moisture with minimal soil disturbance. Crop rotation is combined with herbicides and a fallow period in order to conserve moisture. Data collected from 1972 through 1974 showed significant decreases in stalk rots of sorghum under the ecofallow system when compared to conventional tillage. Possible reasons for this include stress reduction due to the presence of mulch and the fact that crop rotation was used. The same crop is never planted back into the stubble and mulch from the previous crop. This prevents the build-up of pathogens which could move into the current crop.

Modification of planting date, row spacing, and plant populations combined with selection of suitable pesticides offer additional means of coping with residues and soil environment changes associated with conservation tillage. These practices, when combined with varieties selected for conservation tillage fields and crop rotation can provide very high levels of protection from common plant disease problems associated with residue.

SUMMARY

Conservation tillage practices may influence plant diseases directly or indirectly. Residues may provide an excellent habitat for pathogens and serve as foci for disease development; provide a nutrition source of fertilizers, chemicals, and related materials; may favor the development of antagonistic microorganisms; may modify the chemical and physical environments of the soil to favor either the host crop or the pathogen; and may produce compounds during decomposition that beneficially or adversely affect the pathogen; may affect the physical distribution of roots to make them resistant or vulnerable to pathogen invasion.

LITERATURE CITED

Boosalis, M.G., B. Doupnik, and G.N. Odvody. 1981. Conservation Tillage in Relation to Plant Diseases. Pages 445-474 in CRC Handbook of Pest Management in Agriculture, Vol. 1. CRC Press, Inc., Boca Raton, FL. 597 pp.

Boosalis, M.G., D.R. Sumner, and A.S. Rao. 1967. Overwintering of conidia of Helminthosporium turcicum on corn residue and in soil in Nebraska. Phytopathol. 57(9):990-996.

Burns, E.E. and M.C. Shurtleff. 1973. Factors affecting overwintering and epidemiology of Helminthosporium maydis in Illinois (abstract). Phytopathol. 62(7):749.

Chaudhary, M.R. and S.S. Prihar. 1974. Root development and growth response of corn following mulching, cultivation, or interrow compaction. Agron. J. 66(2):350-355.

Guenzi, W.D., T.M. McCalla, and F.A. Norstadt. 1967. Presence and persistence of phytotoxic substances in wheat, oat, corn, and sorghum residue. Agron. J. 59(3):163.

Schuster, M.L., W.A. Compton, and B. Hoff. 1972. Reaction of corn inbred lines to the new Nebraska leaf freckles and wilt bacterium. Pl. Dis. Rept. 56(10):863-865.

Snyder, W.C., M.N. Schroth, and T. Christou. 1959. Effect of plant residue on root rot of bean. Phytopathol. 49(11):755-756.

Sumner, D.R., B. Doupnik, and M.G. Boosalis. 1981. Effects of reduced tillage and multiple cropping on plant diseases. Ann. Rev. Phytopathol. 19:167-187.

Weihling, J.L. and A.K. Vidiver. 1967. Report of Holcus leaf spot (Pseudomonas syringae) epidemic in corn. Pl. Dis. Rept. 51(5):396-397.

CHAPTER 13

CONSERVATION TILLAGE AND INSECT CONTROL

Robert F. Ruppel
and
Kathleen M. Sharpe

Department of Entomology
Michigan State University
East Lansing, Michigan 48824

With a few notable exceptions, the numbers of Michigan's field and forage crop insects rise and fall erratically over the years. They damage only scattered fields even when generally abundant. Their abundance is determined by a complex of interactive factors termed "natural control." The crop variety, cultural practices, edaphic features, weather, plant stage and stress, and natural enemies (parasites, predators, and diseases) are all parts of the natural control that reduce insect damage to some fields during some years. This reduction in damage is good for Michigan's agriculture. A special problem of these scattered, sporadic attacks is that many growers are not accustomed to regularly checking their fields for insects and are too often "surprised" by insect damage.

Some exceptions to this benign rule have been the foreign pests that have entered the state--cereal leaf beetle, alfalfa weevil, and European skipper, for examples. The number of these exotic species soared until their natural enemies were also brought in to moderate their populations or, in the case of the European skipper, native natural enemies became effective against them. Another exception has been the corn rootworms, both the native northern and the introduced western corn rootworm, that are now the most serious insect problem in the state. Rootworms are damaging almost exclusively only where corn follows corn without rotation. The rootworm problem has, in effect, been created by the change in cultural practices from rotation to corn monoculture. We have good control measures for the rootworms, but we can expect the problem to continue until cultural practices are changed. This example has made us very concerned about the effects of changed cultural practices on our insect pests.

The concern has already been substantiated in part in conservation tillage programs. The cover and high surface moisture have improved the habitat for slugs in no-till corn and soybean fields, and we are having an increasing problem with these hard-to-control pests. In experiments of the junior author at the Kellogg Biological Station, slug damage was most severe in no-till treatments (Sharpe, 1985). The damage was done quickly and significantly reduced stands in some sites. Slugs are usually a problem only near low, wet areas of conventionally tilled crops. Deer mice, another non-insect pest, have damaged corn seed planted no-till in cover and food of grain stubble (Gregory and Musick, 1976). Deer mice have never been a problem under regular tillage. We even had severe damage from lesser clover leaf weevil to soybeans planted no-till in old clover stands. This weevil has not been of concern for years. Soybeans are not even listed among its usual hosts, and it apparently ate the seedling soybeans for lack of other food.

The greatest problem with no-till and insects is the increasing damage to corn from insects associated with grass weeds. The stalk borer, usually found only in the margins and grassy areas of conventionally tilled fields, is increasing in direct proportion to the weed grasses. The hop vine borer and potato stem borer, two corn pests also associated with grasses, are increasing and spreading; they have not been officially found in Michigan as yet. We did find the lined stalk borer while looking for the other two borers, however. This is another old fashioned pest and is as closely associated with grass weeds as are the other borers. The stalk borer complex is hard to control and has been proposed for study by the members of the NC-105 Committee--a joint research group of Agricultural Experiment Stations. Armyworm, primarily a grass feeder, can be a real problem in any field with grass weeds, and no-till fields have been especially vulnerable (Gregory and Musick, 1976). The practice of using rye, another excellent host for armyworm, as a winter cover has also resulted in bad experiences with early infestations of armyworm. There are a number of other corn pests associated with grasses--flea beetles, thrips, and billbugs, for example. Will these also increase in no-till corn?

There is a positive side to the effects of no-till on insects. All and Gallaher (1977) found that lesser cornstalk borer damage was greatly reduced in no-till as compared to conventionally tilled corn in Georgia. The increased soil moisture in the no-till was the primary factor that suppressed this borer. Cheshire and All (1979) showed that alternative food sources in the no-till system were also partially responsible for the reduction in damage. Reduced till in Nebraska also affected the distribution of western corn rootworm eggs and retarded the emergence of the adults sufficiently to reduce their damage to grain set in corn (Pruess et al., 1968). There may, therefore, still be a chance to reduce rootworms through cultural measures (Chiang, 1973; Chiang et al., 1971). The reduced soil disturbance and increased soil cover in conservation tillage fields that favor some pests should also favor some of their natural enemies (Blumberg and Crossley, 1983). Could the increased cover favor predaceous ground beetles to the point that they effectively control cutworms, armyworms, and stalk borers? The great emphasis in studies of tillage systems and insects has been on the soil insect complex. Tillage practices affect the plants that, in turn, could affect the foliage insects. The junior author, for example, has strong indications that the first generation of the European corn borer may be less severe in no-till compared to conventional tillage (Sharpe, 1985). This reduction may be caused by the increased failure of the borer to establish on the smaller plants in the no-till plots early in the season.

With hind-sight, we can explain most of the changed pest patterns that have occurred. We never would have projected the deer mouse or lesser clover leaf weevil problems, however, and are uneasy waiting to see what happens next. With their high reproductive rates, changes in insect populations can be levered by only slight changes in survival; for example, an insect that produces 100 progeny per individual maintains a stable population if the cumulative mortality is 99 percent and doubles in numbers if the rate is only 98 percent. We must assume that changes in tillage practices will result in changes in habitat, perhaps very subtle, that favor some species and impede others. We can expect such changes to accelerate as a larger proportion of our acreage goes into new tillage types. Even so, the changes may be very slow in developing and as variable as our soils and the tillage practices that alter the soils.

Insects are always most damaging when they are unexpected. We now know of the slug, stalk borer, and others and can alert growers to them. The grass weed problem, now our main concern, will diminish as growers become accustomed to handling it. We must still fear the unknown. There are a large number of insect species now in low numbers in our soils that could increase to damaging levels given the proper circumstances. The seedcorn beetle, for example, is common but not presently a problem in Michigan and of minor concern elsewhere. Will it become a serious problem under conservation tillage? We do not know which species will succeed, and a major effort must be made to anticipate the problems in time to head off any widespread losses. Frankly, we have been reacting to established problems when we should be far enough in advance of them to avoid them.

We do have some things going for us. Studies to find improved chemical means for immediate control of pests are active in many states. These are generally funded by the pesticide industry with emphasis, of course, on the material rather than the pest. More studies funded through neutral sources and directed at control of the pest, independent of the product, would be a real aid. Slug control comes to mind as this is an unusual pest and few chemical companies are interested in it as it presently offers a very limited market. We also have biological studies, such as the NC-105 studies of stalk borer, underway that should in time give us satisfactory long-term solutions to our problems. Both the control and biological studies are directed at recognized problems, however, and not purposely aimed at detecting incipient problems.

Small plot studies, including those of the junior author, on the effects of conservation tillage systems on the insect complex (not just soil insects) are also underway in various forms in several states. These could be an excellent means of discovering potential problems. They are extremely important in identifying and measuring parameters that influence population changes, and they are basic research tools. Their limits are that they are financially constrained in their size and numbers; paying to have all possible tillage methods out in all major agricultural areas and then monitoring them for all insects would require negotiations with the World Bank. An alternative would be to have a greater number of cooperators (perhaps extension and governmental program personnel, industry people, and individual growers) monitor conservation tillage plots or fields on a well planned, regular basis. The costs involved with this, we believe, could be repaid the first time that some "surprise" is avoided.

We already have parts of this cooperator network in place. Our normal communications are very good; the detection of lesser clover leaf weevil and lined stalk borer came through these sources last season, for example. The weekly newsletter "Pest Alerts" is the central focus for the information received from these sources in Michigan. We also have the Cooperative Crop Monitoring Service (CCMS) with a large number of cooperators (including pest scouting services) who make regular, uniform reports that are quickly processed at reasonable cost. A program of cooperators reporting through CCMS and aided by a good identification service would be a big aid in heading off problems before too many people get hurt.

Our communications are especially good with persons who have special interests in insect control. Our contacts, understandably, diminish as we go away from our specific field. We have already seen that one thing leads to another; stalk borers increase as grass weeds increase as no-till fields increase. An understanding among the different disciplines at all levels is needed to assure that, at the worst, we do not simply pass problems down the line. This present workshop is one means of aiding the needed understanding, and we hope that they will be continued. The real secret to understanding is frequent, continuous contacts, and we hope to be in touch with you more than we have been in the past.

SUMMARY

Problems with slugs, deer mice and lesser clover leaf weevil have been detected in fields of conservation tillage, and problems with armyworm and the stalk borer complex have increased where grass weeds have been a problem in such fields in Michigan. Changes in tillage practices can be expected to cause similar problems with other insects and related organisms over time. We have generally reacted to insect problems as they have appeared, however, while the need is for programs directed at detecting incipient problems in time to devise means of avoiding them or minimizing their effects. It is very evident that communication and cooperation among all disciplines are needed to assure sound development of conservation tillage. Those include:

(1)   Special studies of immediate control for the recognized pests should be funded independently of industry-supported research.

(2)   Biological studies of these pests should be supported as sources of fully satisfactory means of managing them.

(3) The major emphasis should be on detecting incipient problems and, hopefully, finding means of avoiding them. Toward this end:

(a)   Small plot studies of the effects of tillage systems on insects should be supported.

(b)   Cooperators should be encouraged to check conservation tillage fields and to report their findings.

(c)   A formal program of monitoring conservation tillage plots or fields should be established with support of the Cooperative Crop Monitoring Service and an identification section and with the aid of cooperating entities and individuals.

(d)   Communication at all levels of the different disciplines should be increased as an essential means of satisfactorily avoiding or minimizing problems.

LITERATURE CITED

All, J.N. and R.N. Gallaher. 1977. Detrimental impact on no-tillage corn cropping systems involving insecticides, hybrids, and irrigation on lesser cornstalk borer infestations. J. Econ. Entom. 70(3):361-365.

Blumberg, A.Y. and D.A. Crossley, Jr. 1983. Comparison of soil surface arthropod populations in conventional tillage, no-tillage and old field systems. Agro-Ecosystems. 8:247-253.

Cheshire, J.M., Jr. and J.N. All. 1977. Feeding behavior of lesser cornstalk borer larvae in simulations of no-tillage, mulched conventional tillage, and conventional tillage corn cropping systems. Environ. Entom. 8(2):261-264.

Chiang, H.C. 1973. Bionomics of the northern and western corn rootworms. Ann. Rev. Entom. 18:47-73.

Chiang, H.C., D. Rasmussen, and R. Gorder. 1971. Survival of corn rootworm under minimum tillage conditions. J. Econ. Entom. 64(6):1576-1577

Gregory, W.W. and G.J. Musick. 1976. Insect management in reduced tillage systems. Ohio Agricultural Research and Development Center (O.A.R.D.C.), Journal Article No. 81-76.

Pruess, K.N., G.T. Weekman, and B.R. Somerhalder. 1968. Western corn rootworm egg distribution and adult emergence under two corn tillage systems. J. Econ. Entom. 61(5):1424-1427.

Sharpe, K. 1985. The influences of tillage and other cultural practices on invertebrate damage and numbers in field corn. MS. Thesis, Dept. of Entomology, Michigan State University, East Lansing, MI.

CHAPTER 14

EFFECTS OF PARAQUAT AND ATRAZINE ON NON-TARGET SOIL ARTHROPODS

Richard J. Snider
Department of Zoology
Michigan State University
East Lansing, Michigan 48824

John C. Moore
Natural Resource Ecology Laboratory
Colorado State University
Fort Collins, Colorado 80523

Jusup Subagja
Faculty of Biology
Gadjah Mada University
Yogyakarta, Indonesia

INTRODUCTION

It is accepted that no-tillage (conservation tillage) strongly relies on the action of herbicides to be successful. The judicious application of this class of chemicals can result in high production yields. In fact, we have replaced the plow and cultivator with herbicides. One consequence of this practice is that a pedestrian attitude has developed about their use. In 1975 Chevron Chemical Company published a bulletin entitled, "Getting Started with No-till." In it, practical advice for using paraquat in no-tillage farming is presented. While careful instructions are given for applying the herbicide, no mention is made about the toxicity of the compound. On the back page one finds a warning that paraquat should be handled with care. Only recently have we publicly become aware that paraquat is potentially dangerous to animal life.

Another chemical commonly used in no-tillage systems is atrazine. Galston (1979) has pointed out that this compound has been found to disrupt chromosomes in plants. If we couple the toxicity problem with possible herbicide carryover, a potential threat to production and to a beneficial fauna may exist. The explanation of herbicide carryover is reported by Smika and Sharman (1982).

Research on possible herbicidal effects on non-target organisms was begun in our laboratory at Michigan State University during the 1970s. These studies began when field work focused on tillage effects and soil animal response. Our purpose here is to report some results of research conducted in experimental plots in and the laboratory on soil animals that comprise the mesofauna. Also included are some remarks about the direction and needs of current research.

The soil fauna may be viewed in two ways. Certain species are economically important because they are pests. Others are beneficial; they process litter, increase soil pore space, prey on pest species, or are an important food source for other organisms. Some, indeed, are unknown to us.

For the last several decades the soil fauna have come under close scrutiny. In the past only individual phyla, orders and even individual genera received the attention of specialists. We are now at the point where ecosystem analysis is no longer looked upon with apprehension. Large groups such as nematodes, protozoans, annelids, and arthropods are recognized as indicators of soil condition, not merely as academic curiousities.

Early reports about arthropods in agroecosystems usually stated that their populations temporarily decreased because of tillage practices (Tischler, 1955). It was recognized that destruction of soil structure affected soil animal populations. Review of cultivation effects upon soil animals may be found in Butcher et al. (1971) and Ghilarov (1975).

The practice of no-tillage farming results in virtually minimum disturbance of ground cover and soil structure. This, in turn, directly affects soil animals. These advantages may be counteracted by the use of chemical compounds to control weeds and animal pests. Changes in the soil fauna are expected when no-tillage methods are introduced. These changes should be understood since agro-chemicals may initiate long term effects (Edwards, 1969; Edwards and Thompson, 1973).

In our laboratory, one taxon that is often used in experiments is the springtails (order Collembola). Herbicide studies utilizing springtails have shown very different effects. First of all, in the field, effects may be masked by environmental factors such as pore space, moisture, temperature, solar radiation, chemical residues and pH. Of several herbicides generally applied, only those of the triazine group were consistent in reducing springtail populations, at least temporarily (Fox, 1964; Edwards, 1970; Popvici et al., 1977). Paraquat produced contradictory results. According to Curry (1970), springtail populations decreased, while Edwards (1975) found their numbers increased.

Laboratory tests are helpful in understanding and interpreting direct herbicide effects. Eijsackers (1975; 1978a; 1978b) studied the effects of 2,4,5-T on springtail under controlled conditions. Two species were studied by Subagja and Snider (1981) in an experiment that included continuous exposure to paraquat and atrazine. They reported effects on reproduction, instar duration and mobility. Still, while we can document such effects in the laboratory, their magnitude in the field is poorly understood. Many investigators have tried to separate chemical and tillage effects, but thus far we have results that are less than perfect.

MATERIALS AND METHODS

A field study area was set up on the campus of Michigan State University. Through the efforts and cooperation of the Departments of Zoology and Crops and Soil Sciences, a grassy field was prepared for conventional and no-tillage corn production. Each plot was divided into a 6 x 15 m rectangle. The following treatments were each replicated four times: (1) Moldboard plow, corn, no herbicide; (2) Moldboard plow, corn, cultivate, no herbicide; (3) Moldboard plow, corn, cultivate, atrazine; (4) No-tillage, corn, atrazine; (5) No-tillage, corn, atrazine and paraquat; (6) No-tillage, corn, paraquat; (7) No-tillage, corn, no herbicide; and (8) Grassland plot.

Plot preparation and planting followed methods commonly used by area farmers. Irrigation was provided during dry periods. On May 18 and June 12, 0.47 and 0.94 cc/m2 of atrazine was broadcast in the treatment plots. On May 18, 116.9 cc/m2 of paraquat was applied. Sampling for soil arthropods began on May 6 and was conducted twice a month, except for September, November, and December, when one sample period was utilized. The sampling device was a metal corer with a diameter of 5 cm. Each sample was taken to a depth of 10 cm. Five random soil cores were taken from each plot. The cores were divided into two subsamples. The samples were placed in plastic bags and transported in a cool chest to prevent mortality before extraction. The Tullgren funnel method was used for extraction. Animals were collected in 95 percent ethanol and later sorted and counted to species. In addition, on each sample date, the soil pH, moisture, temperature and area climatological data was taken.

Effects of Paraquat

The springtail fauna was selected as an indicator organism because of its abundance and easy identification. The total population of springtails decreased temporarily after application of paraquat. The effect could be detected one month after application and lasted about two months. Several species were selected for analysis: Hypogastrura manubrialis Tullberg; Brachystomella parvula (Schaffer); Onychiurus armatus (Tullberg); Tullbergia granulata Mills; Isotoma notabilis Schaffer; Lepidocyrtus pallidus (Reuter); and Pseudosinella violenta (Folsom). Of these species, only B. parvula, L. pallidus and T. granulata had significantly decreased.

Data collected on age structure, based on body length classes, supported the conclusion reached in laboratory studies that paraquat affects reproduction in these animals. Newly hatched juveniles of B. parvula were not found for two months after application of paraquat. These are animals that reproduce about every 30 days. They were, however, found in plots without herbicide application. The same applied to T. granulata.

Lepidocyrtus pallidus was drastically reduced, with recruitments not appearing for over 4 months. In springtail community analysis, while the abundance of some species was reduced, composition was not altered. Eventually those species whose numbers were reduced, recovered. This does suggest that continuous no-tillage practice with paraquat may not deleteriously affect springtail communities. It also points out the recuperative power of these animals.

Laboratory Studies on Paraquat and Atrazine

The results in this section are reported in detail in Subagja and Snider (1981). Two species of springtail, Folsomia candida (Willem) and Tullbergia granulata Mills, were selected for laboratory study. Our purpose was to subject them to concentrations of paraquat and atrazine similar to those used in no-tillage farming.

The animals were reared in culture containers, described by Snider et al. (1969), at 15.5°C in constant darkness. Brewer's yeast was supplied as food. Eggs were selected from stock cultures and transferred to new containers, one egg per container. Starting on the day of hatching, individuals were fed yeast impregnated with either paraquat or atrazine. Concentrations were: 600, 1000, and 5000 mg/kg. In the field, the standard concentrations are 600 mg/kg for paraquat and 500 mg/kg for atrazine. Daily observations of 20 replicates for each concentration ended after 22 weeks.

The following summarizes the results. Based on replicates of treatment and control, mortality for F. candida at the termination of the experiment was 5 percent for control, 0 percent for 600 mg/kg, 10 percent for 1000 mg/kg, and 5 percent for 5000 mg/kg paraquat. Mortality was 5 percent for 600 mg/kg, 10 percent for 1000 mg/kg, and 5 percent for 5000 mg/kg atrazine. Mortality for T. granulata was 0 percent for control, 0 percent for 600 and 1000 mg/kg, and 5 percent for 5000 mg/kg paraquat. Terminal mortalities were 0 percent for 600 and 1000 mg/kg, and 55 percent for 5000 mg/kg atrazine.

In cultures with 5000 mg/kg paraquat or atrazine, the individuals were smaller than in either of the controls or the other two concentrations. Egg production, likewise, was lower at 5000 mg/kg for both compounds.

These results were obtained by continuous exposure to paraquat or atrazine after hatching. Similar conditions cannot be found in nature. However, they are suggestive and support the field study conclusions reported earlier. In the laboratory individuals were observed to be repelled by heavily treated yeast (5000 mg/kg atrazine). They may have starved to death rather than eat adulterated food.

Studies by Loring et al. (1981) point out that T. granulata was found in lower numbers in no-tillage corn where atrazine and paraquat were used. They suggest that this might be related to selective feeding by the springtails. Eijsackers (1975; 1978b) speculated that in nature, springtails may be able to avoid contaminated food.

CONCLUSIONS

The investigations described here have pointed out the importance of knowing what some of the interactions of an agroecosystem are. In a recent paper, Crossley et al. (1984) state that mutualistic interactions are characteristic of ecosystems with certain self-regulating capacities. They suggest that no-tillage research may hold the potential for identifying positive interactions within the soil. Plant pathologists believe new systems experience a "grace period" of 3 to 5 years before imbalances show up at levels that need correction. Therefore, long-term tests over 6 to 7 years are needed.

Clearly there are effects produced by the two herbicides, paraquat and atrazine. Many, if not all, soil animals respond to these and other perturbations in an identifiable and quantifiable manner. We are on the threshold of learning how the delicate balance between organisms and agrochemicals can be carefully adjusted for production advantages without wholesale deleterious effects on non-target species.

In no-tillage systems the role of the soil arthropods largely focuses on litter reduction. They feed upon microorganisms growing on the litter substrate. To get their preferred diet, they must also ingest large quantities of substrate. Once food has passed out of the gut system other microorganisms may colonize the fecal pellets. In the process of organic breakdown and nutrient cycling, these animals act as conversion accelerators.

Awareness of the important role these animals play in nutrient regeneration is now emerging. Stinner and Crossley (1980), reporting on their no-tillage research, suggest that the function of the soil system is mediated by soil biological communities and, further, that the more intact soils of no-tillage systems should be more conservative of nutrients.

Finally, we close with a quote from Wallwork (1983), who has energetically raised the role of the fauna to new heights. "Quite properly, the biogeochemical cycling of carbon and nitrogen has been given a central place in the thinking of ecologists interested in nutrient fluxes in soil ecosystems. Until quite recently, it was considered that these fluxes were very largely regulated by the activities of soil microflora. It is now becoming apparent that this microflora can only operate efficiently within the limits imposed by the microfauna."

SUMMARY

This paper reported the results of field and laboratory studies on the effects of paraquat and atrazine on soil arthropods. Springtails were subjected to diets in the laboratory that approximated field doses for paraquat and atrazine. Lethal effects for both compounds were at 5000 mg/kg. Field experiments demonstrated an effect on at least 3 species of springtail that lasted up to 2 months. It is pointed out that soil animals are important in soil interactions and that perturbations by chemical means may seriously affect their stability.

LITERATURE CITED

Anon. 1975. Getting started with no-till. Cheveron Chemical Co., Ortho Division, San Francisco, Calif. Prog. No. 76168-02. 12pp.

Butcher, J.W., R.M. Snider, and R.J. Snider. 1971. Bioecology of edaphic Collembola and Acarine. Ann. Rev. Ent. 16:249-288.

Crossley, D.A., Jr., G.J. House, R.M. Snider, R.J. Snider, and B.R. Stinner. 1984. The positive interactions in agroecosystems. In: Agricultural Ecosystems: Unifying Concepts. G. House, R. Lowrance & B. Stinner (eds.). John Wiley & Sons, Inc., New York, pp 73-81.

Curry, J.P. 1970. The effects of different methods of new sward establishment and the effects of the herbicides paraquat and dalapon on the soil fauna. Pedobiol. 10:329-361.

Edwards, C.A. 1969. Soil pollutants and soil animals. Sci. Amer. 220:88-99.

Edwards, C.A. 1970. Effects of herbicides on the soil fauna. Proc. 10th Brit. Weed Control Conf. :1052-1062.

Edwards, C.A. 1975. Effects of direct drilling on the soil fauna. Outlook on Agriculture. 8:243-244.

Edwards, C.A. and A.R. Thompson. 1973. Pesticides and the soil fauna. Residue Rev. 45:1-79.

Eijsackers, H. 1975. Effects of herbicides 2,4,5-T on Onychiurus quadriocellatus Gisin (Coll.). In: Progress in Soil Zoology. J. Vanek (ed.). Academia Prague:481-488.

Eijsackers, H. 1978a. Side effects of the herbicide 2,4,5-T on reproduction, food consumption, and moulting of springtail *Onychiurus quadriocellatus* Gisin (Collembola). A. Ang. Ent. 85:341-360.

Eijsackers. H. 1978b. Side effects of the herbicide 2,4,5-T affecting mobility and mortality of springtail *Onychiurus quadriocellatus* Gisin (Collembola). A. Ang. Ent. 86:349-372.

Fox, C.J.B. 1964. The effects of five herbicides on the number of certain invertebrate animals in grassland soil. Can. J. Plant Sci. 44:405-409.

Galston, A.W. 1979. Herbicides: a mixed blessing. Bioscience 29:85-90.

Ghilarov, M.S. 1975. General trends of changes in soil animal populations of arable land. In: Progress in Soil Zoology. J. Vanek (ed.). Academia, Prague:31-39.

Loring, S.J., R.J. Snider, and L.S. Robertson. 1981. The effects of three tillage practices on Collembola and Acarina populations. Pedobiol. 22:172-184.

Popovici, I., G. Stan, D. Stefan, R. Tomescu, A. Duncan, T. Tarta, and F. Dan. 1977. The influence of atrazine on soil fauna. Pedobiol. 17:209-215.

Smika, E.D. and E.D. Sharman. 1982. Atrazine carryover and its soil factor relationships to no-tillage and minimum tillage follow-winter wheat cropping in the central great plains. Colo. State Univ. Exp. Sta., Bull. 144. 4p.

Snider, R.J., J.H. Shaddy, and J.W. Butcher, 1969. Some laboratory techniques for rearing soil arthropods. Mich. Ent. (1):357-362.

Stinner, B.R. and D.A. Crossley, Jr. 1980. Comparison of mineral element cycling under till and no-till practices: an experimental approach to agroecosystems analysis. In: Soil Biology as Related to Land Use Practices, D. Dindal (ed.). U.S. EPA, Washington, DC, pp. 280-288.

Subagja, J. and R.J. Snider. 1981. The side effects of the herbicides atrazine and paraquat upon *Folsomia candida* and *Tullbergia granulata* (Insecta, Collembola). Pedobiol. (22):141-152.

Tischler, W. 1955. Effects of agricultural practice on soil fauna. In: Soil Zoology. D. Devan (ed.). Butterworth's, London. 215-230.

Wallwork, J.A. 1983. Soil fauna and mineral cycling. In: New Trends in Soil Biology. Proc. VIII Inter. Colloq. Soil Zool. P. Lebrum, et al. (eds). Dieu-Brichart, Louvain-La-Neuve. 29-33.

PART III

THE ECONOMICS AND ENERGY REQUIREMENTS FOR
SELECTED CONSERVATION TILLAGE CROP SYSTEMS

CHAPTER 15

DATA REQUIRED FOR ECONOMIC EVALUATION:
SHORT AND LONG RUN

Anthony M. Grano[1]
Economic Research Service
U.S. Department of Agriculture
East Lansing, Michigan 48823

INTRODUCTION

Economic information and evaluations of resource management systems applicable to conservation decisions are needed by farmers, agencies and technicians concerned with soil conservation, and decision makers involved with policy and program implementation. What alternative resource management systems are available for erosion control; what effect will they have on erosion; and what will be their costs and benefits? This information is needed if farm owners and operators, technicians and policymakers are to make wise and informed decisions concerning the investment of private and public funds for soil conservation.

Data Required - Short Run

Due to characteristics of the erosion problem and potential alternative solutions, data needs for economic analysis and evaluation are interdisciplinary in nature. Data inputs for economic evaluation can be catagorized as follows: (1) soils, (2) soil erosion factors for the Universal Soil Loss Equation (USLE), (3) land use, such as crops and/or rotations, (4) tillage methods, (5) conservation practices and costs, (6) production inputs and costs,

[1]Current address: Economic Research Service; U.S. Department of Agriculture; GHI Building, Room 428; 500 12th Street, N.W.; Washington, DC 20250

(7) yields and (8) product prices. Each of these data sets become inputs to various methods for analyzing and evaluating the economic impacts of alternative resource management systems for erosion control. Resource management systems are combinations of crops and/or rotations, tillage methods, and conservation practices.

Depending on the intended use of the information, soils data may be very specific or at various levels of aggregation. For planning and analysis at the farm or field level, soil mapping units may be the suitable unit. For studies dealing with watersheds, river basins, or larger regional areas, aggregations of soil mapping units or even aggregations of land capability subclasses are used. Generally, aggregation of soils has attempted to group soils that are relatively homogeneous with respect to productivity, management, production inputs, and erosion treatments. It is also important to note that many current studies are concerned with identifying those soils that are potentially erosive and/or currently eroding at excessive rates to target their analyses. Concentrating the analyses and evaluations on the soils that represent a high percentage of the erosion problem (either protentially or currently) is essential.

The Universal Soil Loss Equation (USLE) has been used extensively by various studies to estimate gross annual sheet and rill erosion. The factors for this equation are obtained from a variety of sources and technical personnel for each soil. However, some factors (such as the C factor for cover) must be obtained for each alternative tillage method, crops and/or rotations, etc.

Crops and/or rotations, tillage methods, and conservation practices have an effect on rates of erosion. Consequently, alternative resource management systems for erosion control need to first specify relevant land uses, crops and/or rotations that are feasible on the various soils. Inputs of this nature need to come from agronomists, crop specialists and/or soil scientists.

Tillage methods are the second component of the alternative resource management systems. Commonly used tillage methods are conventional tillage (fall or spring), reduced or minimum tillage, and zero tillage. Similar to the issue of crop feasibility on various soils, technical specialists must determine the feasibility of various tillages on each soil or soil grouping.

The third component of the resource management systems for erosion control are conservation practices such as terraces, contour farming, grass waterways, etc. Annual capital and maintenance costs need to be estimated for each of these practices that are feasible alternatives on various soils.

Production inputs and costs need to be estimated for each crop and/or rotation, tillage system, and conservation practice combination by soil or soil group. The quantity and costs of these inputs are for items such as seed, chemicals, fertilizers, fuel, machinery, custom work, drying, labor, etc. Production cost and net income estimates are developed utilizing various crop budget computer systems available from many universities and government agencies, such as the Soil Conservation Service (SCS), Economic Research Service, and Extension Service. Data for these inputs and costs come from a broad spectrum of disciplines and sources--agricultural engineers, agronomists, crop and soil scientists, farm management specialists, district conservationists, extension personnel, etc.

The prices of all commodities (crops) are required for analysis of resource management systems. These commodity prices times yields, minus production and conservation practice costs, will determine the net income (returns) for each alternative resource management system by soil or soil group. Also, for each resource management system by soil or soil group, gross annual sheet and rill erosion can be estimated by use of the USLE. Consequently, for each soil or soil grouping, net income and gross erosion can be estimated for the specified alternative resource management systems (crop/rotations - tillage - conservation practice).

Study Examples - Short Run

Numerous studies have been made throughout the nation that have utilized the data inputs specified above for analyzing and evaluating the economic and erosion inpacts of alternative resource management systems. These studies have presented information in a variety of forms--such as by soil on a per hectare basis and/or per farm, by soil grouping for watershed or river basin, etc. Some examples of these study results will serve to illustrate how the data inputs and outputs were utilized to present various economic evaluations of alternative resource management systems.

Table 1 presents the partial results of a study by Doster, et al. (1983). Cost and returns per hectare for various tillage systems on a 304 ha corn/soybean farm on light colored (low organic matter), often highly erodible,

Table 1.   Cost and returns per ha hypothetical 304-ha, corn/ soybean farm, group III soil.

	Fall Plow	Fall Chisel	Spring Plow	Spring Disk	Till Plant	No-Till
	- - - - - - - - - kg/ha - - - - - - - -					
Yield						
Corn	5394	5645	5645	5645	5896	5896
Soybean	1742	1804	1804	1804	1928	1928
	- - - - - - - - - - $/ha - - - - - - - - -					
Value						
Corn @ $3.00/ 25.45 kg						
Soybeans @ $7.00/ 27.27 kg	570	588	588	588	615	615
Production Costs						
Fertilizer/Seed	96	96	96	96	96	96
Pesticides	42	42	42	44	62	84
Fuel, Repairs, Drying	84	84	84	77	74	74
Part-time Labor	15	12	12	10	2	2
Credit Miscell.	27	27	27	27	27	27
Overhead Costs						
Machinery	94	91	111	91	77	72
Crop Storage	27	30	30	30	32	32
Family Labor	67	67	67	67	67	67
Land Rent	67	67	67	67	67	67
Returns	42	72	49	79	111	94

sloping, well drained soils (group III) are estimated. Till plant and no-till had the highest net returns, while fall plow and spring plow were the least profitable systems. Yields of corn and soybeans for these soils were higher than for other tillage systems.   Fuel, part-time labor and machinery costs were lower for till-plant and no-till, while pesticide costs were higher.   Estimates of gross erosion for each tillage system were not made.   For other soil groupings, results from this study displayed varying implications of the profitability of various tillage systems.

The Central Ohio River Basin Study conducted by the U.S. Department of Agriculture (1983) analyzed alternative resource management systems by soil for various counties. Table 2 presents comparative net returns and soil loss for numerous crop management alternatives (combinations of rotations and tillage systems) for a soil. The table also displays soil saved per hectare in tons per year compared to the base management alternative of corn-soybeans fall plowed. The last column is the predicted change in net return compared to the base management alternative. The average annual net return per hectare excluding land cost, that farmer would receive with average management levels and yields for corn-soybeans, fall plow is about $68, and the soil loss is about 72 tons per hectare per year (see astericked rotation in Table 2). There are numerous crop management alternatives that result in higher net returns and lower soil loss than the corn-soybean fall plow alternative. All of the crop management systems that have a soil loss below T (tolerance level) have higher net returns than the base management alternative.

The use of the data inputs and outputs to analyze a land treatment program for a watershed and/or river basin area can be demonstrated from a study of the Upper West Branch Pectonica Watershed in Wisconsin. The U.S. Department of Agriculture (1981) study analyzed and evaluated the impacts of various levels and combinations of rotations, tillages, and conservation practices. Table 3 summarizes the plan elements and impacts in a generalized form. However, each alternative has various combinations of resource management systems (rotations, tillages and conservation practices) imbedded in the analyses. Alternative 3, which emphasizes conservation tillage and no-till, contour and contour stripcropping, results in the highest net returns to cropland acres and lowest soil loss. Each plan also displays the uses of fertilizer, pesticides and fuel. Even though a limited number of alternatives were analyzed, the results of the analyses indicate that increasing conservation tillage and no-till acreages have beneficial economic and soil loss impacts.

These three studies were cited to illustrate various methods of analyzing data inputs and presenting outputs that can provide useful economic information to farmers, planners, and policy makers concerning the short-run costs and benefits of alternative resource management systems. For individual soils, farms and regions; economic and physical information is presented to assist in selecting cost-effective resource management systems. There are many other studies that could also have been cited to demonstrate the uses of the data inputs and outputs.

Table 2. Soil management alternatives and comparative net returns--Pickaway County, Ohio, 1982.

| Crop Management Alternatives | | | | Mgt. Alternatives *Compared with C-SB Fall Plow | |
Rotation	Tillage	Soil Loss MT/ha/yr	Net Return Per ha	Soil Saved MT/ha/yr	Change Nt Ret. Per ha
- - - - - - - - - - - - - - Over T - - - - - - - - - - - - - -					
C-C-C	Chisel Disk	20.21	178.46	52.22	110.29
C-SB	No-Till	13.47	177.32	58.95	109.15
C-SB-SB	No-Till	23.60	156.03	48.85	87.86
C-C-SB	Chisel Disk	28.65	150.72	43.80	82.55
C-SB-C-SB-WX	No-Till	11.79	142.27	60.64	74.10
C-SB	Chisel Disk	37.07	136.86	35.38	68.69
C-C-C-M-M-M	Chisel Disk	11.79	134.52	60.64	66.34
C-C-C	Fall Plow	67.37	123.87	5.05	55.70
C-C-C	Spring Plow	60.64	123.87	11.79	55.70
C-SB-SB	Chisel Disk	43.80	123.01	28.65	54.83
SB-SB-SB	No-Till	33.70	113.45	38.75	45.28
C-C-SB-W-M-M	Chisel Disk	18.52	111.62	53.90	43.45
C-SB-C-SB-WX	Chisel Disk	23.60	109.92	48.85	41.74
C-C-C-M-M-M	Fall Plow	25.28	107.22	47.17	39.05
C-C-C-M-M-M	Spring Plow	21.89	107.22	50.54	39.05
C-C-M-M-M	Fall Plow	16.84	103.89	55.59	35.72
C-C-M-M-M	Spring Plow	13.47	103.89	58.95	35.72
C-C-SB	Spring Plow	62.32	101.94	10.10	33.76
SB-SB-SB	Chisel Plow	57.27	95.27	15.15	27.10
C-SB-WX	Chisel Plow	28.65	91.93	43.80	23.76
C-SB	Spring Plow	64.01	90.97	8.42	22.80
C-C-SB-W-M-M	Spring Plow	23.60	87.24	48.85	19.07
C-C-SB	Fall Plow	70.76	86.75	1.68	18.57
C-SB-SB	Spring Plow	65.69	80.00	6.74	11.83
C-C-SB-W-M-M	Fall Plow	28.65	79.63	43.80	11.46
C-SB-C-SB-WX	Spring Plow	50.54	73.19	21.89	5.01
*C-SB	Fall Plow	72.45	68.17	0.00	0.00
C-SB-WX	Spring Plow	38.75	61.35	-6.20	-0.44
SB-SB-SB	Spring Plow	69.06	58.07	3.37	-10.10
C-SB-C-SB-WX	Fall Plow	57.27	54.96	15.15	-13.21
C-SB-SB	Fall Plow	74.13	49.60	-1.68	-18.57
C-SB-WX	Fall Plow	45.48	46.14	26.96	-22.03
SB-SB-SB	Fall Plow	75.81	12.47	-3.37	-55.70
- - - - - - - - - - - - - - Under T - - - - - - - - - - - - - -					
C-C-C	No-Till	5.05	241.20	67.37	173.02
C-C-SB	No-Till	10.10	198.61	62.32	130.44
C-C-C-M-M-M	No-Till	4.38	165.89	68.05	97.71
C-C-M-M-M	No-Till	3.88	150.82	68.56	82.65
C-C-SB-W-M-M	No-Till	5.05	135.58	67.37	67.41
C-C-M-M-M	Chisel Disk	10.10	125.72	62.32	57.55
C-SB-WX	No-Till	6.74	118.91	65.69	50.73

Soil = Alexandria Silt Loam; Slope = C 6 to 12; R = 150; K = 0.37; L = 186; S = 8.0; T = 5.0; Soil Symbol AdC2

However, one recent study is especially noteworthy because of the impacts it has had on other studies and agency evaluation procedures. Raitt (1981, 1983) developed and presented a computerized system for estimating and displaying short-run costs and benefits of soil conservation systems for specific soils. The system utilizes the same types of data inputs as discussed above. Erosion rates, costs per hectare and costs per ton reduction are displayed in a schematic diagram that permits one to observe the cumulative effects of adding practices to an initial practice. Combinations of practices are ranked by cost per ton reduction and cost per hectare. The model also computes the effects of incremental changes in underlying conservation input cost on per acre practice costs. These procedures and displays are currently being used extensively by the Soil Conservation Service in their incremental analysis of land treatment in projects.

Data Required - Long Run

Recently, there has been increased interest in determining the long-term physical effects of erosion (soil depletion) and the resulting economic consequences. The type of data required for economic analysis and evaluation are somewhat similar to those required for the short run. Emphasis in soil depletion studies have generally concentrated on long run estimates of changes in several major factors: (1) productivity (yields), (2) fertilizer use and (3) energy (fuel) use. However, depending on the models and methods used to estimate changes in these major factors, extensive amounts of data may be required.

Many soils have a shallow layer of topsoil that may have already been depleted to the point where their productivity is very low or they cannot be used for row crop production. Other soils with deeper topsoils and productive subsoils have been kept productive by substituting increased amounts of fertilizer. Power requirements (fuel) for various tillage operations are affected by various soil characteristics such as texture, structure, organic matter, bulk density, etc. Increased costs and reduced incomes due to losses in productivity and needs for increased inputs have not always been recognized and/or accounted for.

Currently, there are numerous studies underway and models being developed for estimating the long-term effects of continued erosion on productivity, etc. Once of the recent studies in this area of work was the U.S. Department of Agriculture (1980) study of soil depletion in the Southern Iowa Rivers Basin. The primary purpose of this study

Table 3. Erosion and sediment control cropland: comparison of project alternatives, Upper West Branch Pecatonica, Wisconsin.

Plan Elements and Impacts	Units	Alternatives				
		1	2	3	4	5
1. Conventional Tillage	Hectares	0	0	0	0	15,660
2. Conservation Tillage 909–1364 kg residue	Hectares	5,387	5,387	2,910	5,387	46,392
3. Conservation Tillage 1364–1818 kg residue	Hectares	72,499	72,499	52,339	71,497	14,477
4. No-Till	Hectares	---	---	22,638	1,003	1,359
5. Terraces	Hectares	395	20,019	---	3,004	3,379
6. Contour Stripcropping	Hectares	52,480	52,480	52,480	52,339	17,100
7. Contour	Hectares	20,019	20,019	21,595	3,984	8,776
8. Grass Waterways	Hectares	205	205	205	205	153
9. Diversions	Meters	3,048	3,048	3,048	3,048	2,286
10. Cropland Receiving Treatment	Hectares	77,887	77,887	77,887	77,887	36,314
11. Total Area	Hectares	77,887	77,887	77,887	77,887	77,887
12. Soil Loss	MT/ha/yr	4.45	3.79	2.96	5.32	12.15
13. Soil Loss	MT/yr	56,726	48,476	37,785	67,990	155,230
14. Sediment Delivered	MT/yr	5,559	4,751	3,703	6,619	15,212
15. Cropland Treatment Impacts						
a. Total net Return[1]	$/yr	2,877,000	2,045,700	3,606,500	2,741,900	2,790,000
b. Fertilizer Use						
1. Nitrogen	MT/yr	1,008	1,018	1,001	1,001	1,001
2. Phosphorus	MT/yr	910	847	770	869	846
3. Potassium	MT/yr	1,826	1,904	2,000	1,694	1,674

Table 3. Erosion and sediment control cropland: comparison of project alternatives, Upper West Branch Pecatonica, Wisconsin (continued).

Plan Elements and Impacts	Units	Alternatives				
		1	2	3	4	5
15. Cropland Treatment Impacts (Cont.)						
c. Pesticide Use						
1. Corn Insecticide	kg/yr	386,760	339,240	348,920	376,640	332,860
2. Hay Insecticide	l/yr	59,251	56,411	59,062	56,411	46,379
3. Corn Herbicide	l/yr	93,514	85,564	113,959	88,782	92,378
d. Fuel Use						
1. Gasoline	l/yr	381,150	384,178	395,911	355,412	330,052
2. Diesel	l/yr	1,265,326	1,214,985	1,317,937	1,176,000	1,138,150

[1]Data are for cropland only and reflect only the erosion control practices of the program.

Table 4. Harvested cropland by erosion phase projected over time*, Southern Iowa Rivers Basin.

Erosion Phase	1974	2000	2020
		Hectares	
1	277	231	203
2	493	408	311
3	79	210	335
Total	849	849	849

*Assuming 1974 land use and treatment to be constant.

was to predict the effects that current rates of soil erosion, if continued, will have on individual soils in Southern Iowa by the year 2020. The study is an evaluation of the increasing cost of erosion, over a long period of time, due to reduced productivity and increased fertilizer and fuel costs.

Table 4 presents area of harvested cropland by erosion phase over time. If depletion were to continue at current rates, the distribution of harvested cropland among erosion phases will change over time. There will be more hectares of soils in erosion phase 3 in the year 2020 and less hectares in erosion phases 1 and 2. Table 5 presents the specified annual cost of soil depletion for the Southern Iowa Rivers Basin for the years 2000 and 2020. With generally

Table 5. Specified annual cost of soil depletion*, Southern Iowa Rivers Basin.

	Million Dollars	
	Year	
Item	2000	2020
Fertilizer	2.0	2.6
Yields	4.3	8.1
Fuel	0.4	0.7
Total	6.7	11.4

*Assuming 1974 conditions throughout time period to 2020.

decreasing yields as soils change from one erosion phase to another, increased fertilizer inputs and additional fuel required to till depleted soils, annual costs of soil depletion were estimated to be nearly $7 million in the year 2000 and over $11 million by 2020.

Another model recently developed by Williams, et al. (1983) to determine the relationship between soil erosion and soil productivity is the Erosion-Productivity Impact Calculator (EPIC). The EPIC model consists of series submodels: hydrology, weather, erosion, nutrients, plant growth, soil temperature, tillage and economics. While EPIC a is fairly comprehensive model, it has been developed specifically for application to the erosion-productivity problem. Output from the EPIC model will input to the Iowa State University linear programming model and be utilized to analyze alternative erosion control alternatives for the national RCA program. EPIC will have potential uses beyond the RCA anaylsis--research, project level planning and national planning and policy studies.

The University of Illinois has developed a model for analyzing the long-run physical and economic implication of resource management systems (Elveld , 1983). The research, supported by the SCS, is designed to provide SCS field staff with a tool to assist farmers in conservation planning. The model analyzes the physical and economic trade-offs of alternative resource management systems for farms, and it provides information and guides to policymakers for setting conservation targets and subsidies.

Needs for Further Research and Data

Much information and data has been developed and analyzed nationally and locally on the physical and economic dimensions of erosion and alternative methods for controlling erosion (resource management systems). However, substantial increases in research to improve and/or develop data and methods are necessary.

One major area for additional effort is on the relationship of yields and soil erosion. More research data is needed to establish the effects that continued erosion (soil depletion) will have on crop productivity for specific soils and locations. The relationships between soil depletion and plant nutrients, chemicals and energy use need to be better understood.

Estimates of yields are an important data input for physical and economic evaluations, both short and long term. How yields are impacted by various rotations, tillage methods and conservation practices on various soils is another area of research needing more attention. This area of research is currently receiving increased attention, but considerably more effort is required.

More information is needed on the cost of various conservation practices (terracing, contour farming, grass waterways, etc.). Estimates of initial construction costs, annual capital and maintenance costs, and loss of area used for cropping are needed for specific situations (soils, fields, etc.). also, the relationships of conservation practices to efficiency of various field operations need to be better understood.

Cost data related to conversion from one tillage system to another tillage system are needed. Switching from one tillage system to another involves investments in new machinery or attachments, modifications to existing equipment, and/or utilizing machinery compliments that may not be economical. Also, new technology relating to tillage methods, crop varieties, conservation practices, etc., must be incorporated into data bases for analysis and evaluation.

As pointed out earlier, the data inputs required for physical and economic analysis and evaluation are interdisciplinary in nature. Close cooperation and coordination is needed between physical and social scientists. Team approaches are an essential ingredient to comprehensive analyses and evaluation of erosion, its impacts and potential solutions to the problem. Because of the nature of the problem, data requirements, methods of analyses and close cooperation are needed among Federal, State and local agencies and groups, universities, etc.

SUMMARY

Data inputs required for short and long run economic evaluation of alternative resource management systems to control erosion are presented. Data inputs were catagorized as follows: (1) soils, (2) soil erosion factors, (3) land use (crops and/or rotations), (4) tillage methods, (5) conservation practices and costs, (6) production inputs and costs, (7) yields and (8) product prices. Examples of various studies are displayed and discussed to illustrate the results and use of dta inputs and outputs for economic evaluation fo resource management systems. Needs for further research, additional data and interdisciplinary cooperation are presented.

LITERATURE CITED

Doster, D.H., D.R. Griffith, J.V. Mannering, and S.D. Parsons. 1983. Economic returns from alternative corn and soybean tillage systems in Indiana. J. Soil Water Conser. 38(6):504-508.

Elveld, B., G.V. Johnson, and R.G. Dumsday. 1983. SOILEC: Simulating the economics of soil conservation. J. Soil Water Consrv. 38(5):387-389.

Raitt, D.D. 1981. COSTS: Selecting cost-effective soil conservation practices. J. Soil Water Conserv. 38(5):384-386.

U.S. Department of Agriculture. 1983. Pickaway, soil and water conservation district resources inventory, 12.

U.S. Department of Agriculture. 1980. Soil depletion study, reference report: Southern Iowa Rivers Basin. 18 pp.

U.S. Department of Agriculture. 1981. Upper West Branch Pecatonica Watershed. 46-47.

Williams, J.R., K.G. Renard, and P.T. Dyke. 1983. EPIC: A new method for assessing erosion's effect on soil productivity. J. Soil Water Conserv. 38(5):381-383.

CHAPTER 16
MACHINERY REQUIREMENTS AND COST COMPARISONS
ACROSS TILLAGE SYSTEMS

C. Alan Rotz
Agricultural Engineer, USDA/ARS
Agricultural Engineering Department

J. Roy Black
Professor of Agricultural Economics
Agricultural Economics Department

Michigan State University
East Lansing, Michigan 48824

INTRODUCTION

Concern has developed in recent years over the decrease in water quality and soil productivity which has resulted in the Saginaw Bay area of Michigan from excessive erosion. This concern has prompted the Agricultural Conservation Program administered by the Agricultural Stabilization and Conservation Service to promote, through a cost share program, conservation tillage practices which reduce erosion and the associated pollution. Adoption of conservation tillage practices must continue after the cost share program has terminated. Voluntary adoption by farmers was uncertain because of the lack of knowledge of the economic impacts of such practices to the farmer. A study was undertaken to do a comparative economic analysis of the new conservation tillage systems to the tillage practices traditionally used in Michigan's Saginaw Bay Watershed.

Machinery requirement is a major component in an economic comparison of tillage systems. An initial survey of equipment used on farms in the area showed that machines were often not properly matched to one another nor to the available power on the farm. Farmers normally do not buy a complete, well-matched set of machines at one point in time. Instead they buy equipment when needed and attempt to match the new equipment to their current situation. As farm size and cultural practices vary, the farmer will likely not have a "best" set at any point in time, even though, it may be the most "economical" for him at that time.

To make a proper comparison between tillage systems, a properly sized machinery complement was needed for the different tillage systems in different soils under various crop rotations. Selecting machinery for farms which produce several crops is a complex problem. Interactions among machines and between the machinery subsystem, land and weather create a problem which requires a systems approach for solution. Because of the large number of variables and relationships, a computer model is required to obtain definitive answers.

Several computer models have been used for machinery selection. Whole farm, profit maximizing linear programs have been used in applied research, extension and farm machinery company workshops (Doster, 1981; Black and Harsch, 1976). Linear programming for machinery selection, however, has limitations as a solution algorithm. Either mixed integer linear programming must be used, which is very expensive, or a conditional optimization approach must be used where the user provides input for a specific complement, deduces the consequences and provides input or a revised complement in an interactive procedure until a suitable complement is determined.

Least cost models have been used to calculate the total cost for machinery systems for a particular crop or crop rotation as a means for selecting the "best" machinery complement. Hunt (1983) developed a method for determining the least cost size for a given machine used for a particular crop area to minimize the total cost of the machine including the cost of timeliness. The least cost approach has been used to calculate the cost of machinery systems of various sizes to determine the least cost system for corn and soybean rotations (Burrows and Siemens, 1974). A limitation of the least cost approach is that it cannot match equipment; it can only select among complements of matched equipment.

Models have been developed which select machinery based upon time constraints of various operations on the farm. Hughes and Holtman (1976) described an algorithm of this type which selected and matched machine implements and power sources for a multiple crop farm. The time constraint algorithm was further developed by Singh (1978) and Wolak (1981). This approach has worked well in determining the "best" size for a machinery complement given the date constraints, suitable days available and operation requirements. A limitation is that the cost of timeliness is not considered. Even though a given complement can satisfy the constraints, a larger complement may provide a more economical set when crop losses are considered.

Our study was undertaken with the objective of extending the development of a time constraint algorithm to enable machinery selection for conventional and conservation tillage systems as a basis for economic comparison. The algorithm was modified to include a cost analysis with timeliness considered so that a machinery complement was selected for maximum profit on a given farm. Tillage and planting equipment for various crop rotations on different soils were compared.

A MACHINERY SELECTION ALGORITHM

Methods of Selection

Two basic methods for the selection of farm machinery were integrated to form the selection algorithm. These methods were: (1) a capacity and power match; and (2) a cost analysis. This was done by selecting several machinery systems which were properly matched to meet the imposed time constraints and then to select the least cost system based upon the cost analysis.

Capacity and power matching of farm machinery have been described in detail by Hunt (1983). With this method, machines are first sized to complete the job in the time alloted to each operation. The capacity of each machine is determined as the area which must be covered by that operation divided by the time available. When the required capacity is known, the machine size required can be determined, given the field speed and efficiency of the operation.

Capacity matching is complicated by the fact that many operations are interrelated by time. When two operations are done in sequence, time devoted to one operation takes time away from the other. Time, therefore, must be properly divided between sequential operations. Two operations may also be done parallel to one another, i.e. they must complete two different functions within the same time. This also links the capacities of the selected machinery.

Farm machinery is also often interrelated or linked by size. This is most true for row equipment. As an example, if a 12-row planter is required on the farm, then the combine must be either a 6-row or 12-row unit.

Properly sized machinery should also match the power available on the farm as well as possible. Implements cannot be sized larger than the associated tractor can pull. Likewise, implements should not be undersized because this causes inefficient use of the tractor. Tractors must be selected to match the power requirements of all implements as closely as possible, allowing only for some reserved power.

Power requirements again involve an interrelationship between machines. If two implements are pulled by the same tractor, then their power requirements should be similar. This again places a constraint on the implement size selected.

With the use of a computer, relationships for machine capacity and power requirements can be properly balanced to provide a well-matched machinery system. Often several good systems can be found for the same farm. When several well-matched systems are functionally suitable for a given farm, which one should be selected as the "best" set? At this point a cost analysis can be useful to select the "least cost" or "maximum profit" system.

A cost analysis is done by summing the costs of owning and operating a system of machines over a period of time, normally its useful life (Rotz et al., 1981). The time value of money should be considered in a cost analysis. Money or cash transactions in the future will not have the same value as the identical money or cash transactions today. Both inflation and the discount rate of the farm enterprise affect the value of money through time. During periods when certain costs are experiencing a high inflation rate, some systems can be affected more than others, which will influence the decision of the "best" machinery system.

Another component in a cost analysis which may influence machinery selection is the tax incentive system. Certain deductions or credits may influence certain machinery systems more than others, which again will influence the decision process.

As the second major step in the selection process, therefore, a cost analysis was done to calculate total system costs for each well-matched set of machines. Total cost was determined as the sum of all costs of owning and operating the system plus the cost due to timeliness. The timeliness cost was defined as the loss in crop value experienced because the job was not completed within an

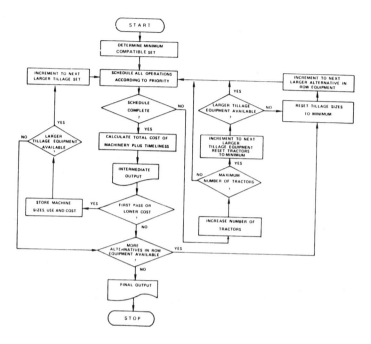

Figure 1. Algorithm used for machinery selection based upon time constraints and total system cost.

optimum time.   Consideration of the timeliness cost allows the selection of the maximum profit system.   In some cases it is less expensive overall to allow some crop loss and to own a smaller machinery set.   In most cases, though, it is least expensive to own larger machinery and avoid all crop loss.

Machinery can be selected by either the capacity and power matching method or the cost analysis method alone.   A better solution can be determined, however, with more efficient use of the computer by using a procedure which combines both methods.

A Machinery Selection Model

A computer model was developed which combined the capacity and power matching and cost analysis methods for the selection of farm machinery.   A diagram of the information flow through the computer algorithm is shown in Figure 1.   The major steps of the model through the decision process will be discussed.

Required minimum capacities for each machine were calculated based upon the operations required, the area to be covered by each operation and time constraints for the operation, all of which were inputs to the model. In the first step of the algorithm, the minimum capacity needed to complete each field operation within the time interval specified was determined. Based upon the minimum capacities, a set of combines and implements was selected from commercially available sizes. Tractor sizes were determined based upon power requirements for the implements. Implements with large power requirements were assigned to the primary tractor while all others were assigned to the secondary tractor. The match between row equipment was checked. The size of the planter, row cultivator and ammonia applicator were set either equal or double the size of the combine.

After the minimum-sized set of equipment was selected, all operations were scheduled according to the hours available with first priority given to harvest. If any operations were to be done through custom hire, they were scheduled accordingly. If a problem was encountered in completing any of the operations within the given time constraints, the number of tractors available was increased. If that did not allow completion, the size of the tillage equipment was increased. If the schedule still could not be completed, each post tillage implement was independently increased until all operations were satisfactorily completed.

When the scheduling of all operations was complete, the total annual cost of owning and operating the set of machines was determined including custom hire and the cost of timeliness. If any crop was lost due to untimely operations, the value of the loss was added to the cost of the machinery set.

After the first set of machines and the associated costs were determined, a check was made to determine if a lower cost (including timeliness) set of equipment could be determined. Implement or combine sizes were incremented and the entire process was repeated. The planter, row cultivator and anhydrous ammonia applicator were independently doubled in size. Within each of these major steps the tillage equipment was also incremented. When a lower cost system was not found they were returned to their original size. The combine size was then increased to the next larger size and the process was repeated. All machinery complements which fulfilled all operations were listed in the output from the smallest to the largest.

The least cost (maximum profit) set of machines was selected as the "best".

INFORMATION REQUIREMENTS FOR MACHINERY SELECTION

Like many other computer studies, machinery selection is only as good as the information fed into the process. Much information is required, some of which has a rather large impact on the machinery set and some of which has a more minor effect. Major blocks of information required by the machinery selection model will be discussed in relation to the availability and accuracy of data and the sensitivity of the model to changes in the data. This discussion is provided to give scientists more direction as experiments are designed for the collection of data on tillage systems.

Farm Parameters

Parameters which describe the farm include: size or total land area, crop rotation and predominant soil type. This information would normally be readily available for a given farm and would be specified to the model as input.

A computer algorithm prefers information in a very organized form which places an additional constraint on crop rotation. The total farm must be divided into equal-sized parcels of land for each crop in the rotation. This allows the same crop rotation year after year. On a real farm, the farmer often does not follow a regular rotation but fluctuates crop areas from year to year. To represent the real farm, the crop area input must be set as close as possible to the actual crop areas but yet maintain equal-sized parcels in a complete rotation.

Predominant soil type is required in the estimation of suitable days for field work and implement power requirements. The soil type specified as input must, therefore, correspond to the soil type used in estimating the suitable days and power requirements. Our approach was to assume three soil types: coarse (sandy), medium (loam) and fine (clay). A better description of soil texture would improve the accuracy of the model only in that it would allow a more accurate description of suitable days and power requirements if this corresponding data were available.

Crop and Weather Parameters

Crop and weather parameters include data on timeliness cost and the suitable work time available for various operations. The timeliness penalty set in the model for crop loss due to untimely operations may have a small effect on machinery selection. Normally the model will select to avoid all timeliness cost; therefore, the magnitude of the cost will not have much effect, at least not within a reasonable range of error.

The timeliness period can have a large affect on machinery selection. Since the model generally prefers to avoid all timeliness costs, setting a narrow period for zero timeliness cost will essentially force the model to select based upon a short time constraint, i.e. very large equipment will be selected. For this reason, care must be taken in setting the timeliness periods so that they are as realistic as possible. Also, operations must always be given a starting date which allows the operation enough time for completion before a timeliness cost period is encountered. A set of data on timeliness costs for various crops has been prepared by scientists at Michigan State University (Table 1).

Table 1. Timeliness costs for planting and harvesting operations.

Crop	Planting		Harvesting	
	Loss %/day	Penalty $/ha/wk	Loss %/day	Penalty $/ha/wk
Corn	1.0 after May 13	43	1.0 after Nov 18	43
Wheat	1.0 after Sep 30	31	0.5 after Jul 29	31
Oats	2.4 after Apr 22	43	0.5 after Aug 26	10
Soybean	1.0 after May 20	35	1.0 before Oct 1, after Oct 15	35
Navy Bean	0.7 before Jun 4, after Jun 24	35	0.7 before Sep 3, after Sep 17	35
Sugar Beet	1.0 after May 6 3.0 after May 13	70 87	--- ---	-- --

Source: Unpublished research data, Crops and Soils Department, Michigan State University.

Suitable time available throughout the working season is required for machinery selection. This type of information is generally hard to obtain because of the long period of time required to obtain such data. Actual records have been kept for several years at a few locations across the U.S. When available, this type of data will provide a good basis for selection. Another source for suitable-day data is computer simulation. A model developed by Rosenberg et al. (1982) can be used to generate suitable-day probabilities from weather data for a specified location. The modeling approach provides a fast and relatively accurate means for obtaining suitable work-day data. Suitable day data should be specified as a function of the type of operation, the week of the year, the soil type and the probability level. A probability level of 0.8 or 8 years out of 10 normally provides the best basis for selection.

Accuracy of the suitable work-day data is of moderate importance. A ten percent error in the estimation of suitable time will cause, at most, a ten percent error in machine size selected. It is possible under some conditions that an error in suitable-day data will not influence the "maximum profit" machinery set. This occurs when a larger set of machines is more economical than a smaller set which forces the model to select the larger set regardless of the time available for the work.

Machine Parameters

Parameters which describe the machinery include: commercial sizes available, field efficiency, field speed and power requirements. The model selects machines from commercially available sizes specified in the model. If sizes are improperly listed, of course an error will occur. Data on commercially available sizes is readily available from manufacturers of farm equipment.

Field efficiency and speed are used by the model to determine the capacity of machines. Both have a similar effect on machinery selection. Accuracy for this information is of moderate importance in that a ten percent error will cause an error in machine selection of ten percent or less. Data on field efficiency and speed are available in farm machinery management publications (Hunt, 1983; Hahn et al., 1984).

Of all machine parameters, the most difficult data to obtain are draft or power requirements for various machines. This is also the most critical information. Power requirement information is available from many sources; however, no one source provides data for a wide range of machines in similar or "standard" conditions. Data must be pulled together for one location where a comparison between machines or across soil types provides consistent differences. When data is simply pulled together from many sources, the conditions behind the numbers vary widely, which leads to inconsistency. When data is not consistent across machines and soil types, unrealistic results can be obtained from the machinery selection model.

Error in the power requirement data will directly affect the tractor size which will in turn indirectly affect other implement sizes. Because tractors are the highest cost component of the machinery system, a small error in tractor size can cause a considerable shift in the set of machinery selected. For this reason, power data should be specified as accurately as possible in the model. The best source for power data is field measurement of the data at the location under study. Field measurement is expensive, but necessary if one is to select valid machinery systems.

A group of scientists at Michigan State University have developed a set of power requirement data. Data from all sources was drawn together across machine operations and soil types. Typical (or median) values were selected for each machine on each soil type. Any values which seemed inconsistent when compared with the others were adjusted. The results are given in Table 2.

Economic Parameters

A cost analysis of machinery requires much input information. In terms of machinery selection, the most important parameter is probably initial cost. Again, consistency across machine types is most critical. All machine prices must be given in terms of the same dollars, i.e., 1984 dollars. A good source of this information is commercial suppliers of agricultural equipment. In taking a survey of machine prices, one must always be careful to work with the same price, say list price. Prices can vary considerably as influenced by different financial arrangements.

Tax benefits can have a major effect on the machinery selected. When income tax deductions and investment credits are considered, machinery costs are reduced substantially. Tax benefits may favor some machines more than others which will shift the selection to a different set of machines. If one does not want to complicate machinery selection with tax considerations, the tax rate can be set to zero.

Table 2. Tractor power requirements (kW/meter of width) for operating various field machinery on three major types of soil.

| Machine Type | ---- Soil Type ---- | | |
	Coarse	Medium	Fine
Bean Puller	4.8	5.7	6.6
Beet Topper	9.5	9.5	9.5
Beet Lifter	18.9	22.3	22.3
Subsoiler	20.4	25.0	31.1
Fertilizer Spreader	2.3	2.3	2.3
Chisel Plow	21.4	22.7	24.1
Moldboard Plow	18.9	26.1	28.6
Light Disk Harrow	8.0	9.3	10.0
Heavy Disk Harrow	16.8	18.2	19.5
Field Cultivator	10.4	11.4	12.5
Grain Drill	3.2	4.1	4.1
Row Planter	6.1	6.6	7.3
No-Till Planter	6.6	7.0	7.7
Sprayer	2.7	2.7	2.7
Row Cultivator	3.6	4.3	5.0
Ammonia Applicator	9.3	10.2	11.4

Interest, discount and inflation rates should always be set in relation to one another. Under a stable economy, these factors will not have much effect on machinery selection. When inflation is greater for one component, say fuel, in relation to another component, such as labor, these parameters can influence selection. To simplify the selection procedure, the discount and inflation rates can be set equal so that money does not change value over time.

A cost analysis of agricultural machinery also requires parameters to describe the remaining value of the machines at the end of the period of analysis and the repair costs over the period of analysis. Values for these parameters are given in the literature (Hahn et al., 1984). An error or small change in those values would normally have little effect on machinery selection.

Fuel and labor prices are required in the cost analysis. Changes in these prices may have some effect on machinery selection. In general, though, small changes in these prices will have little or no effect on machinery selected for maximum profit of the farm.

A detailed sensitivity analysis was previously reported for the model (Rotz et al., 1984). The analysis showed that as the probability level of suitable weather was decreased, smaller machinery complements became feasible; however, little change occurred to the economically optimum system. Changing from a clay to a sandy soil allowed for a smaller machinery set. Larger farm sizes allowed machinery to be used more efficiently for a lower cost per unit area. The model was further verified by demonstrating a reasonably close comparison between actual and simulated machinery sets for two Michigan farms.

MACHINERY COMPARISONS

The machinery selection model has been used to determine machinery requirements for a variety of farm sizes and crop rotations for both conventional and conservation tillage systems. Several examples will be discussed for corn, navy bean, soybean, sugar beet, wheat and alfalfa rotations on 200 and 500 ha farms. Weather and crop parameters were set for Eastern Michigan conditions which included a fine (clay) soil.

Machinery requirements and costs for the five selected crop rotations are listed in Tables 3 through 7. In all cases, the conservation tillage machinery can be owned and operated at a lower cost than conventional tillage machinery. The saving in machinery cost through conservation tillage varies across rotations from a high of $30/ha to a low of $11/ha. The reduction in machinery cost obtained through conservation tillage occurs primarily because fewer machines are required. In general, smaller machines are also required with conservation tillage which again contributes to the lower cost. Smaller equipment is possible because with conservation tillage there is less competition for available time between implements.

There are two major factors which affect the machinery requirements and costs as crop rotations are compared. These factors are closely related. First, the amount of area which must be covered by each operation influences machinery requirements. If a single crop is grown on the farm, then the whole farm area must be covered within a given time constraint which requires larger equipment. As more crops are added, the farm is divided into smaller blocks of land which are worked at different times. Smaller areas done within the same time constraint require smaller equipment. This can be seen by comparing the machinery requirements and costs of conventional tillage given in Tables 3, 4, 5 and 6. The all corn farm places the greatest constraint on time in the early spring. As more crops

Table 3. Machinery requirements and costs for conventional and conservation tillage practices on an all-corn farm in Eastern Michigan.

Machinery	Size Units	200 ha				500 ha			
		Conventional		Conservation		Conventional		Conservation	
		Size	Use(h)	Size	Use(h)	Size	Use(h)	Size	Use(h)
Primary Tractor	kw	76	371	67	319	180	409	142	376
Secondary Tractor	kw	43	196	46	185	86	276	91	262
Combine	row	6	111	6	111	12	139	12	138
Moldboard Plow	bot.	5	157	–	–	12	164	–	–
Chisel Plow	m	–	–	2.1	150	–	–	4.6	175
Field Cultivator	m	4.7	121	3.8	75	11.1	129	8.4	85
Row Planter	row	6	76	6	89	12	95	12	111
Sprayer	row	–	–	12	70	–	–	24	87
Row Cultivator	row	6	94	–	–	12	117	–	–
Fertilizer Spreader	m	12.2	26	12.2	26	12.2	64	12.2	64
Ammonia Applicator	row	6	93	6	93	12	116	12	116
Machinery Cost ($/ha)		187.37		173.30		144.81		133.52	
Fuel Cost ($/ha)		29.75		25.06		32.34		25.94	
Labor Cost ($/ha)		26.09		23.64		12.68		11.95	
Timeliness Cost ($/ha)		0.00		0.00		6.62		0.00	
Total Cost ($/ha)		243.21		222.00		196.45		171.41	
Difference in Total Cost ($/ha)		21.21				25.04			

Table 4. Machinery requirements and costs for conventional and conservation tillage practices on a corn and soybean farm in Eastern Michigan.

Machinery	Size Units	200 ha				500 ha			
		Conventional		Conservation		Conventional		Conservation	
		Size	Use(h)	Size	Use(h)	Size	Use(h)	Size	Use(h)
Primary Tractor	kw	76	295	67	272	135	405	134	342
Secondary Tractor	kw	43	183	46	154	86	244	91	209
Combine	row	6	111	6	111	12	139	12	139
Moldboard Plow	bot.	5	157	–	–	9	219	–	–
Chisel Plow	m	–	–	2.1	150	–	–	4.3	188
Field Cultivator	m	4.7	91	3.8	75	8.4	128	7.5	96
Row Planter	row	6	76	6	89	12	95	12	112
Sprayer	row	–	–	12	52	–	–	24	65
Row Cultivator	row	6	94	–	–	12	117	–	–
Fertilizer Spreader	m	12.2	13	12.2	13	12.2	32	12.2	32
Ammonia Applicator	row	6	46	6	46	12	58	12	58
Machinery Cost ($/ha)		186.12		177.26		137.78		130.98	
Fuel Cost ($/ha)		20.25		16.80		20.55		17.65	
Labor Cost ($/ha)		22.65		20.69		12.13		10.61	
Timeliness Cost ($/ha)		2.95		2.95		5.81		5.81	
Total Cost ($/ha)		231.97		217.70		176.27		165.05	
Difference in Total Cost ($/ha)		14.27				11.22			

Table 5. Machinery requirements and costs for conventional and conservation tillage practices in a corn-corn-navy bean rotation grown in Eastern Michigan.

Machinery	Size Units	200 ha Conventional Size	Use(h)	200 ha Conservation Size	Use(h)	500 ha Conventional Size	Use(h)	500 ha Conservation Size	Use(h)
Primary Tractor	kw	67	387	67	288	135	478	114	445
Secondary Tractor	kw	43	244	46	222	57(2)	228	609(2)	207
Combine	row	6	111	6	111	8	208	8	208
Bean Puller	row	4	57	4	47	6(2)	47	6(2)	47
Moldboard Plow	bot.	4	199	—	—	9	220	—	—
Chisel Plow	m	—	—	2.1	151	—	—	3.7	220
Field Cultivator	m	3.8	126	3.8	75	8.4	143	6.6	109
Row Planter	row	6	76	6	90	8	143	8	168
Sprayer	row	—	—	12	58	—	—	16	109
Row Cultivator	row	6	94	—	—	8	176	—	—
Fertilizer Spreader	m	12.2	17	12.2	17	12.2	43	12.2	43
Ammonia Applicator	row	6	62	6	62	8	116	8	116
Machinery Cost ($/ha)		188.80		185.21		144.49		134.47	
Fuel Cost ($/ha)		24.80		19.98		25.11		20.73	
Labor Cost ($/ha)		28.57		23.90		17.59		16.43	
Timeliness Cost ($/ha)		0.00		0.00		1.41		3.94	
Total Cost ($/ha)		242.17		229.08		188.60		175.57	
Difference in Total Cost ($/ha)		13.09				13.03			

Table 6. Machinery requirements and costs for conventional and conservation tillage practices in a corn-navy bean-wheat-sugar beet rotation grown in Eastern Michigan.

Machinery	Size Units	200 ha				500 ha			
		Conventional		Conservation		Conventional		Conservation	
		Size	Use(h)	Size	Use(h)	Size	Use(h)	Size	Use(h)
Primary Tractor	kw	88	275	88	226	106	499	95	448
Secondary Tractor	kw	38	385	38	324	57	505	61	448
Combine	row	4	124	4	124	8	156	8	156
Bean Puller	row	2	85	2	85	6	71	6	71
Beet Lifter	row	4	32	4	32	4	81	4	81
Beet Topper	row	4	28	4	28	4	71	4	71
Moldboard Plow	bot.	5	117	—	—	7	212	—	—
Chisel Plow	m	—	—	2.4	98	—	—	3	197
Field Cultivator	m	4.7	91	4.7	60	6.6	164	5.6	127
Grain Drill	m	4	20	4	20	6.1	33	4	51
Row Planter	row	4	85	4	100	8	107	8	125
Sprayer	row	8	13	8	78	16	16	16	98
Row Cultivator	row	4	140	—	—	8	176	—	—
Fertilizer Spreader	m	12.2	13	12.2	13	12.2	32	12.2	32
Ammonia Applicator	row	4	35	4	35	8	43	8	43
Machinery Cost ($/ha)		220.31		214.98		134.27		123.79	
Fuel Cost ($/ha)		29.16		24.00		22.46		19.55	
Labor Cost ($/ha)		30.17		25.95		17.86		16.18	
Timeliness Cost ($/ha)		2.45		0.00		1.19		0.32	
Total Cost ($/ha)		282.09		264.93		175.78		159.84	
Difference in Total Cost ($/ha)		17.16				15.94			

Table 7. Machinery requirements and costs for conventional and conservation tillage practices in an alfalfa–corn–corn–sugar beet rotation grown in Eastern Michigan.

Machinery	Size Units	200 ha Conventional		Conservation		500 ha Conventional		Conservation	
		Size	Use(h)	Size	Use(h)	Size	Use(h)	Size	Use(h)
Primary Tractor	kw	91	287	88	313	180	345	142	342
Secondary Tractor	kw	38	286	38	201	86	284	91	217
Combine	row	4	83	4	83	6	138	6	138
Beet Lifter	row	4	32	4	32	6	54	6	54
Beet Topper	row	4	28	4	28	6	47	6	47
Moldboard Plow	bot.	6	190	—	—	12	137	—	—
Chisel Plow	m	—	—	24	131	—	—	4.6	146
Field Cultivator	m	5.6	76	4.7	78	11.1	96	8.4	85
Grain Drill	m	4.0	7	4.0	7	12.2	6	6.1	11
Row Planter	row	4	85	4	100	12	71	12	84
Sprayer	row	8	13	8	52	24	11	24	13
Row Cultivator	row	4	140	—	—	12	117	—	—
Fertilizer Spreader	m	12.2	13	12.2	13	12.2	32	12.2	32
Ammonia Applicator	row	4	69	4	69	12	58	12	58
Machinery Cost ($/ha)		235.19		226.31		176.98		156.87	
Fuel Cost ($/ha)		27.09		25.92		24.07		18.92	
Labor Cost ($/ha)		25.24		22.97		11.80		10.74	
Timeliness Cost ($/ha)		0.00		0.00		4.14		0.00	
Custom Work ($/ha)		23.49		23.49		23.49		23.49	
Total Cost ($/ha)		311.01		298.69		240.48		210.02	
Difference in Total Cost ($/ha)			12.32				30.46		

are added and corn area is reduced, machinery size and costs decrease. The decrease is greatest for large farms because machinery is used more efficiently (machines are used more at the limit of their capacity) on large farms.

A second factor which influences machinery size is the number of operations which must be completed. Crops such as alfalfa, navy beans and sugar beets require more operations, and several operations must be done within the same time constraint. Competition between operations forces a larger machinery set to be selected. This can be observed by comparing conventional tillage machinery requirements and costs for the rotations in Table 7 (greater number of operations) to those in Tables 3, 4 and 5 (fewer operations).

When conventional and conservation tillage systems are compared, both of the above factors influence each system. Conventional tillage systems are affected more, however, than conservation tillage systems because they require more operations. A rotation such as the alfalfa-corn-corn-sugar beet rotation which requires many machine operations offers more benefit for conservation tillage than a less intense rotation such as the corn-soybean rotation. This is best illustrated by comparing the difference in machinery costs for the two rotations. On a 200 ha farm where there is less demand on each operation, the saving in machinery cost through conservation tillage is about the same for both rotations (Tables 4 and 7). On the larger farm where the machines are used to the limit of their capacity, the saving in machinery cost is $30/ha for the more intense rotation and $11/ha for the less intense rotation.

On the all corn farm, all operations must be completed in a relatively short time which again places a greater demand on machinery. With more intense use of the machinery, the benefit of conservation tillage is greater with a saving of machinery costs of $22/ha on the smaller farm and $25/ha on the larger farm. As the corn area is reduced, there is less competition for time in the early spring so conservation tillage provides a smaller reduction in costs of $11-15/ha.

SUMMARY

A model for selecting farm machinery complements was extended to model conservation tillage as well as conventional tillage on three major soil types and at three levels of probability of suitable weather. The model selects equipment according to time constraints

for the operations and then uses a cost analysis to compare
feasible alternatives for selection of a "best" complement
of machinery. Parameters which are most critical in the
selection process are: 1) the time constraint placed
on operations to avoid crop loss due to untimely operations,
2) the power requirements of various machines, and 3)
the initial cost of the machines.

Conservation tillage machinery can be owned and
operated at a lower cost then conventional tillage machinery.
Fewer machines are required with conservation tillage
which normally reduces costs. Fewer operations in conservation
tillage also allows more time for remaining operations
which can allow smaller machinery sets. The benefit of
conservation tillage is greatest for crop rotations which
require several operations within the same time constraint
because conservation tillage removes some of the operations
and allows smaller equipment for the remaining operations.

LITERATURE CITED

Black, J.R., and S.B. Harsh. 1976. Corn-soybean-wheat
farm planning guide. Telplan 26, Cooperative Extension
Service, Michigan State University, East Lansing, MI.

Burrows, W.C., and J.C. Siemens. 1974. Determination
of optimum machinery for corn-soybean farms. Transactions
of the ASAE. 17(6):1130-1135.

Doster, H. 1981. Top crop farmer planning model. Cooperative
Extension Service, Purdue University, West Lafayette, IN.

Hahn, R.H., M.A. Purschwitz, and E.E. Rosentreter, (eds.).
1984. ASAE Standards 1984. American Society of Agricultural
Engineers, St. Joseph, MI. pp. 156-162.

Hughes, H.A., and J.B. Holtman. 1976. Machinery compliment
selection based on time constraints. Transactions of
the ASAE. 19(5):812-814.

Hunt, D.R. 1983. Farm power and machinery management.
8th ed., Iowa State University Press. Ames, IA.

Muhtar, H. 1982. An economic comparison of conventional
and conservation tillage systems in the South East Saginaw
Bay Coastal Drainage Basin. Ph.D. Thesis in Agricultural
Engineering. Michigan State University, East Lansing,
MI.

Rosenberg, S.E., C.A. Rotz, J.R. Black, and H. Muhtar.
1982. Prediction of suitable days for field work. Paper

No. 82-1032. American Society of Agricultural Engineers, St. Joseph, MI.

Rotz, C.A., J.R. Black, and P. Savoie. 1981. A machinery cost model which deals with inflation. Paper 81-1513. American Society of Agricultural Engineers, St. Joseph, MI.

Rotz, C.A., H.A. Muhtar, and J.R. Black. 1984. A multiple crop machinery selection algorithm. Transactions of the ASAE. 26(6):1644-1649.

Singh, D. 1978. Field machinery system modeling and requirements for selected Michigan cash crop production systems. Ph.D. Thesis in Agricultural Engineering, Michigan State University, East Lansing, MI.

Wolak, F.J. 1981. Development of a field machinery selection model. Ph.D. Thesis in Agricultural Engineering, Michigan State University, East Lansing, MI.

White, R.G. 1977. Matching tractor horsepower and farm implement size. Bulletin E-1152 SF. Cooperative Extension Service, Michigan State University, East Lansing, MI.

CHAPTER 17

RESULTS OF AN ECONOMIC COMPARISON OF CONVENTIONAL
AND CHISEL PLOW CONSERVATION TILLAGE SYSTEMS IN
THE SOUTHEAST SAGINAW COASTAL DRAINAGE BASIN

J. Roy Black
Professor of Agricultural Economics
Agricultural Economics Department

C. Alan Rotz
Agricultural Engineer, USDA/ARS
Agricultural Engineering Department

Don Christenson
Department of Crop and Soil Sciences

Michigan State University
East Lansing, Michigan 48824

INTRODUCTION

Society's concern about the impact of farming practices
on the environment, as reflected in the Rural Clean Water
Act of 1977, has increased the search for voluntary actions
which reduce wind and water pollution and, ultimately,
sedimentation and waterways and lakes. Conservation tillage
methods that are known to reduce pollution are alternatives
to conventional tillage; however, whether their adoption
would be profitable on the naturally poorly drained fine
textured soils of the Michigan Southeast Saginaw Bay Watershed
was not known. Thus, a four-year study was initiated
in the fall of 1979 to compare, from a whole-farm perspective,
the profitability of conservation tillage systems being
encouraged relative to that of traditional practices used
in the watershed.

This study is one component of the East Central
Michigan Planning Development Region Southeast Saginaw

Bay Monitoring Evaluation Project. Other components include estimation of wind erosion losses under alternative tillage/ planting systems, water quality monitoring, water quality modeling and comparison of point versus non-point source pollution abatement.[1]

Objectives of the project were:

1. To review the literature on conservation tillage/planting systems, with emphasis on factors that influence cost and yield.

2. To conduct field investigations under practical farm conditions using paired four-hectare plots, side-by-side, to evaluate the impact of selected conservation tillage/planting systems relative to conventional tillage systems on cost and yield per hectare.

3. To monitor plant growth, weed and insect incidence, and plant disease, given the management practices followed by the cooperating farmers, since these variables may be associated with yield differences between conservation and conventional tillage systems. Also, to measure residue cover in the fall following harvesting, and in the spring.

4. To develop an awareness of changes needed in farming practices to ensure good performance from conservation tillage systems by using the comparison plots. This must include a "learning by doing" component.

5. To develop a machinery selection model to estimate the machinery and labor required, from a whole farm perspective, for alternative tillage/planting systems. Also, to collect data from cooperators on machinery performance.

6. To integrate the results of previous investigations and of the field investigation to determine the impact of alternative tillage systems on costs, returns, and profit--from a whole-farm perspective. Also, to analyze the sensitivity

[1]This study was funded by the East Central Michigan Planning and Development Region and the Michigan State University Agricultural Experiment Station and Cooperative Extension Service.

of profits to differences in yield between conventional and conservation tillage/planting systems.

The objective of this paper is to summarize the results of the project; details are available in Black et al. (1984), Muhtar (1982), and in Rotz and Black, "Machinery Requirements and Cost Comparisons Across Tillage Systems," Chapter 16 of this volume.

Soil management group[2] is one of the primary determinants of the relative crop yield per hectare potential among tillage/planting systems, Griffith and Mannering (1985). The literature[3] indicates yields under no-till methods on naturally poorly drained, fine textured soils are typically reduced relative to those under conventional tillage in the northern corn belt. However, the literature suggests fall chisel plowing leaves adequate residue cover on soils that may have wind or water erosion problems associated with fall moldboard plowing, and results in lesser or no reduction in yield. Thus, given the practical nature of this study, fall chisel plowing was evaluated as a substitute for fall moldboard plowing. It was the method selected by most farmers in the study area under the cost share agricultural conservation program of the Agricultural Stabilization and Conservation Service of the USDA. All cooperating farmers participated in the ASCS ACP program.

The plan of the paper is as follows. The agronomic results will be discussed, followed by discussion of the economic methodology. The concluding section will analyze, from a whole-farm perspective, the comparative economic performance of the chisel plow variant of conservation tillage in comparison with the moldboard plow conventional tillage system.

---

[2]Michigan State University defines soil management groups as groups of soils with similar properties and yield potential. The groups are defined in the context of dominant texture in the upper 1.5 meters of the profile and the natural drainage conditions under which the soils were formed. Numbers were used to define the dominant texture of the profile (from 0 for fine textured to 5 for sands) and the lower-case letters indicate the natural drainage conditions ("a" for well drained to "c" for poorly drained). A 2.5b soil is a somewhat poorly drained loam or silt loam, while a 2.5c is a poorly and very poorly drained loam and silt loam. See Tilman and Mokma (1976) for additional details, including implications for soil erosion control.
[3]A detailed literature review is contained in Black et al. (1984).

AGRONOMIC RESULTS

The evaluation of the effect of tillage/planting systems on crop yield per hectare is not a simple task. Practices such as pest control, form and method of fertilizer placement, and equipment adaptation all influence the yield that results from a particular tillage/planting system. Soil management group and cropping systems are key determinants of the relative yields among various systems. In this study, they were observed with the objective of being able to attempt to understand sources of differences in agronomic and machinery performance among farms if they were to occur.

Agronomic data were collected from four-hectare side-by-side plots on selected[4] farms in the watershed. Measurements included: yield per hectare, grain moisture at harvest, weed and insect control programs and incidence, plant disease and incidence, standard soil tests conducted by management specialists working in the watershed, fertilization program, plant population, plant growth stage across the growing season, and residue cover. Machinery use and performance parameters were measured in 1981. Detailed measurements on soil moisture were made on one farm in 1981. Soils maps were obtained for each site.

Annual machinery use and variable cost were not estimated for individual farms. Instead, a machinery selection model was developed to analyze machinery requirements on a whole-farm basis for different tillage systems, cropping sequences, and farm sizes (Rotz and Black, Chapter 16). Profit maximizing complements were developed for chisel plow conservation and for moldboard plow conventional tillage systems based on performance and on economic criteria including timelines. The information on machinery performance from the sample farms was used in the development of the parameters of the machinery selection model.

The cooperating farmers maintained ultimate responsibility (and control) for all management decisions. The Michigan State University Cooperative Extension Service County Agricultural Agent on the companion education/demonstration

---

[4]Criteria included: (1) participation in the ASCS ACP program and meeting the associated residue requirements; (2) willingness to maintain side-by-side plots and cooperate on measurements; (3) availability of relatively homogenous soils in an eight-hectare block; and (4) dispersion across portions of Tuscola and Huron Counties in the watershed.

Table 1. Mean monthly rainfall and the probability of a level of rainfall larger than the observed level (based upon rainfall for the last 30 years, Caro location).

Month	1980 PPT	Mean	Prob.	1981 PPT	Mean	Prob.	1982 PPT	Mean	Prob.	1983 PPT	Mean	Prob.
June	3.27	3.09	40	2.51	3.09	60	4.87	3.09	13	2.91	3.09	52
July	5.26	2.92	9	3.08	2.92	40	1.84	2.92	73	2.46	2.92	46
Aug.	2.05	2.96	69	5.23	2.96	8	3.30	2.96	35	3.49	2.96	69
Sept.	5.29	2.98	10	7.70	2.98	2	2.76	2.98	48	3.89	2.98	75

The probability distribution function used for this table was the gamma distribution.

Source: Dr. Fred Nurnberger, Michigan Department of Agriculture, Michigan State University.

project and USDA Soil Conservation Service staff provided technical assistance and conducted discussion tours.

Weather

The literature indicates growing season weather is an important determinant of yield differentials associated with alternative tillage systems. Table 1 depicts mean precipitation during the growing season and the probabilities of rainfall greater than the levels observed. The late summer of 1981 was unusually wet; the probability of rainfall larger than the values observed was less than 10 percent. The 1982 growing season was excellent. The warm spring allowed early planting under both conventional and conservation tillage systems. The 1983 growing season began with a wet, cold spring. Corn planting was delayed, and fields in the project area that were spring moldboard plowed were plowed very late. No-till fields were planted prior to, often significantly prior to, spring moldboard plowed fields.

The substantial variation in weather over the course of the project was very useful. It provided a sense of the "robustness" of alternative tillage/planting systems to alternative weather scenarios.

Yields

Corn. The conservation tillage/planting systems used by cooperating farmers included chisel plow, disk, no-till, field cultivator, and till plant (ridge-till). The chisel plow, however, was the primary conservation tillage system used because of the relatively satisfactory performance of the chisel plow in previous research on naturally poorly drained soils such as the Michigan 2.5b-c soil management groups found in the Southeast Saginaw Bay Drainage Basin. In contrast, no-till systems have performed relatively poorly on naturally poorly drained soils, particularly under monoculture. The till-plant system was gaining popularity in Indiana and Ohio at project conception, but was less well suited to the corn, navy bean, sugar beet, wheat and soybean cropping systems found in the project area as contrasted to corn/soybean sequences.

Table 2 depicts the impact of conservation tillage versus moldboard conventional plow tillage on corn and bean yields. Beans are primarily navy beans. The average corn yield reduction across all years and 30 comparisons associated with the chisel plow conservation tillage system was 1.5 percent. If we compare the number of farms with

Table 2. Impact of conservation tillage versus conventional moldboard plow tillage.

Item	Corn*	Beans
1980		
Average increase in yield, %	1.0	
No. increasing	3/9	No
No. with no difference	1/9	Comparisons
No. decreasing	5/9	
1981		
Average increase in yield, %	-4.2	-
No. increasing	1/8	2/5
No. with no difference	0/8	1/5
No. decreasing	7/8	2/5
1982		
Average increase in yield, %	-1.1	-
No. increasing	3/7	1/5
No. with no difference	1/7	3/5
No. decreasing	3/7	1/5
1983		
Average increase in yield, %	-1.8	-
No. increasing	4/6	3/8
No. with no difference	0/6	2/8
No. decreasing	2/6	3/8

*Chisel plow only.

increasing versus those with decreasing yields, the evidence suggests no differences associated with tillage system. When 1981, an unusually wet year, is removed from the sample there are essentially no differences in yield associated with tillage system. This compares with the synthesis by Doster et al. (1983) that suggested for similar soils in Indiana the differences would range from no reduction for a corn-soybean routine rotation up to 3 percent reduction for continuous corn.

The yields associated with the chisel plow systems were not statistically discernably less than those associated with the moldboard plow system.[5]

The number of farmers using the disk conservation tillage system was smaller than farmers using the chisel plow system. Their average yield reduction relative to the moldboard plow tillage system was 4.5 percent, contrasted to 1.5 percent for the chisel plow. The largest reductions were in 1981 and 1982, 7.1 and 8.1 percent, respectively. There was essentially no difference in 1983.

The number of no-till plots was too small for serious consideration. However, yields were lower under the no-till system with the biggest reduction, as expected, on the Michigan 2.5c management group. This is consistent with farmer observations prior to the execution of the study and leads farmers to be less enthusiastic about no-till, relative to chisel plow, until it can be clearly demonstrated that no-till results in similar yields.

Beans. Dry bean and soybean yields are combined in this discussion (Table 2). There are no statistically discernable differences in yields in each of the three years. Further, the number of instances in which yields were lower under conservation tillage than conventional tillage was exactly equal to the number of years yields were higher.

There were two farms for which conservation tillage was consistently better than conventional tillage over the three-year period. Also, there were two farms that exhibited a reduction in yields under conservation tillage during the same three-year period. This tendency of some individuals to consistently have positive results while others consistently have negative results was much stronger in the dry bean plots than in the corn plots, and could not be associated with any of the factors that were measured.

A review of the subjective farmer comments as to the efficacy of conservation tillage methods across all

---

[5] "A paired t" test was used. In field comparison studies of this type, particularly those where farmers maintain management control, a large number of plots are required to make a valid comparison because of the inherent variability in soils and topography in large plots. Also, large numbers are needed to assure effective randomization. The results, however, are representative of the conditions under which farmers operate.

crops is mixed. Many continue to be concerned about the amount of residue. Nevertheless, farmers are moving to the chisel plow system, particularly on their naturally better drained soils. Many of the cooperating farmers indicated they would probably continue fall moldboard plowing their "heaviest ground," but were satisfied with a chisel plow on their "lighter ground."

Moisture Content

Corn. Differences associated with tillage method were inconsequential in 1980, 1982 and 1983. In 1981, the year in which rainfall was at the 98th percentile level in September, corn from the chisel plow conservation plots contained statistically discernably more moisture at harvest than from the moldboard plowed plot--the average difference was 1.4 points.

Beans. There were no differences in the moisture content of dry beans and soybeans at harvest associated with tillage system.

Weed Control and Herbicide Rates

Cooperating farmers participated in an integrated pest management program and MSU's Cooperative Crop Monitoring Project. Rates of application and types of herbicides were equivalent for conservation and conventionally tilled plots; rates and types were not experimental variables. In 1980, there were no differences in weed control attributable to tillage method for the corn and navy bean plots. In 1981, fields tended to be clean and weed free. There were isolated cases of annual or perennial grasses that incurred on conventional as well as conservation tillage plots. The conservation tillage tilled plots tended to have more perennial grasses, while the conventional tillage plots tended to have more annual grasses and broadleaf weeds. In 1982, there appeared to be slightly more weed problems on the conservation plots (seven of fifteen trials). There were no differences in three of fifteen trials. There were no differences in weeds incidence associated with tillage systems in 1983.

Disease

Bacterial blight, white mold, and mosaic virus were the major diseases affecting beans and generally affected both plots equally.

Insect Population

In 1981, increased army worm and corn borer populations were observed for the conservation tillage treatment in some corn fields; however, the populations were not high enough to have an important impact on yields. Army worm populations were present where small grain cover crops were not effectively killed by spring herbicide applications. In other corn fields, and on bean fields, insect populations were not increased where residues were left on the soil surface.

In 1982, corn root worm counts were significant and were higher for conservation tillage on one farm. Root worm and corn borer populations were within acceptable levels in 1982 for other corn farms; there were no differences associated with tillage method. No insecticides were used on the bean fields, and populations of leaf hoppers and Mexican bean beetles stayed within acceptable levels; there were no differences associated with tillage methods. In 1983, there were no differences in insect populations associated with tillage systems.

Crop Diseases

In 1981, for the most part, no crop diseases were observed that could be attributed to differences in tillage systems. In 1982, eye spot was the disease more common to corn; five out of ten corn plots had more problems with eye spot with conservation tillage. The remaining five farms had equal damage to both plots.

Plant Populations

Cooperating farmers held target seeding rates constant across tillage systems. Seeding rate was not an experimental variable. Incidences where final plant population was lower under conservation tillage were frequently associated with planters that did not properly place the seed into the soil, primarily because of residue interference. Other reasons included planter wheel slippage and excessive packing from planter wheel because of higher soil moisture conditions. Farmers alternating to their regular row crop planters found improved seed placement and depth.

There were statistically discernable reductions in corn plant populations associated with conservation tillage in all four years. There were no statistically discernable differences in the bean plant population associated with tillage.

Plant Growth Rates

In 1980, the corn in the conservation till plots grew more slowly in the spring and was 8 to 13 centimeters shorter four weeks after emergence than corn in conventionally tilled plots. However, they recovered as the season progressed. In 1981, few differences were noted in the rate of plant growth between conservation and conventional tillage. In weeks 8 through 10, 17 percent of the farms had conventionally grown corn ahead of conservation grown corn by one-half growth stage. No differences were recorded after week 12.

Corn in the conservation tilled plots grew slower in 1982, resulting in shorter plants with fewer leaves or less even stand in seven of the eight plots with less naturally well drained soils. The conventional and conservation stands grew equally well on the naturally well drained, naturally better drained soils. There were no differences in the plant growth rate associated with tillage system beyond very early season.

Bean plant growth rates were the same under both tillage systems for all seasons.

Harvest Experience

Cooperators in each of the years, particularly in the very wet year of 1981, said it was easier to harvest the conservation tilled plots than the conventional tilled plots when the soil was very moist.

ECONOMIC ANALYSIS FRAMEWORK

The focus of this study is assessing whether an economic advantage exists, from a farmer's perspective, for the adoption of one of the conservation/tillage planting system alternatives as a replacement for the conventional moldboard plow system. Definition of systems boundaries, in terms of both space and time, is the first step. That is, how are we defining the scope of the problem?

The focus, from a spatial perspective, is on-farm since economic advantages are required for voluntary adoption of management practices and "new" technology. Off-farm impacts, though important, are not included.

The focus, from a time perspective, should be the infinite future--effectively, the next 50 to 200 years when the interest rate is taken into account (see, for example, Dasgupta, 1982; Fisher, 1981; Frohberg and Swanson, example, Dasgupta, 1982; Fisher, 1981; Frohberg and Swanson, 1979). The question is: "Is there an economic incentive to voluntarily adopt one of the conservation tillage systems when only on-farm impacts are considered?" This, however, implies that if erosion reduces the intermediate and longer term potential productivity of the soil, then economic analysis must account for the consequences of the resultant reduced future cash flows relative to what they could have been (and, the associated reduction in land values).

Wind and water erosion under conventional tillage are not likely to result in enough erosion to reduce productivity in the intermediate term for the 0 to 2 percent slopes in the Michigan Soil Management Group 2.5 in the Southeast Saginaw Bay Drainage Basin. Thus, erosion impacts were ignored.[6]

Measures of Economic Performance

The principal measure of economic performance to be used is the "annualized" net return to land. Specifically, net return to land is defined as:

Value of Production (price/unit time yield)

- Seed
- Fertilizer and lime (including soil testing)
- Pesticides (herbicides, insecticides, plant disease regulators, and pest management services)
- Annual use and variable costs for drying corn grain
- Annual use and variable costs for machinery
- Labor (hired plus opportunity cost for family and operator)

---

[6]For soils with significant slope, particularly coarser textured soils, reductions in intermediate- and longer-term potential productivity are typically significant. Thus, the economic analysis must account for reduction in relative future net cash flows and land values under conventional moldboard plow tillage systems. Methodology for this analysis is being published as a companion study, using a sandy loam topsoil over a sandy clay B horizon case study. A rolling landscape is assumed. The methodologies of Larson et al. (1983) and Pierce et al. (1983) can be used in the estimation of potential productivity. The analysis of tillage/planting system replacement is critical in this case because the switch being considered includes moldboard plowing to no-till. The "best" machinery system sizes for these systems differ much more substantially than for the moldboard plow to chisel plow switch.

- Marketing
- Overhead and utilities
- Management services
- Taxes (property and income)

The term annual use cost[7] is associated with capital assets which are kept for several years, as contrasted to being "used up" in the year in which they are purchased. Annual use costs are analogous to the accounting concept of depreciation plus interest on investment.

The analysis is conducted in <u>real</u> terms. That does not mean that we assume there will be no changes in the real prices of individual expenditure categories (e.g., machinery, fuel, pesticides or labor), but that changes will be evaluated relative to the changes in the general price level (e.g., GNP price deflator). The real rate of price change is defined as:

Percent change in real price = Percent change in nominal price - Percent change in general price level

---

[7]The term "annualized" costs (and returns and net cash flows) is used when cost streams are not constant over time. Typically, a cash flow (cash outflow in the case of costs) stream is converted to a present value:

$$PV = \sum_{t=1}^{T} cf_t/(1+r)^t$$

where t is the tth year, T is the planning horizon, cf_t is the net cash flow in the tth year, and r is the interest rate. The sum of future values is converted to a present value by asking what a future cash flow is worth today (i.e., how much would I have to invest today to have $1 in period t). Net cash flow in the tth period is cash inflow (receipts) minus cash outflows. The PV is converted to an "annual" value:

$$AV = (r)(PV)/(1-(1+r)^{-T})$$

AV is a "weighted average annual net cash flow. As an analogy, if PV were the cost of purchasing a house, if the entire cost was borrowed and there was no downpayment, then AV would essentially be the annual house payment.

DIVIDE AND CONQUER

The appropriate time frame is the infinite future, but procedures must be found to divide the problem into meaningful and "do-able" components. The method of analysis should never be more complicated and detailed than the minimum necessary to answer the question. A two-stage approach is used in this study.

The analysis begins with a comparison of the annualized net cash flow to land for alternative tillage/planting systems under the assumption that all machinery is purchased new. The cost calculations are based upon economically best[8] machinery sets, given the crop sequence and farm size, for the conventional and for the conservation tillage/ planting options. If the resultant economic analysis shows that there is no advantage to make the switch from conventional to conservation tillage, then there is obviously no point in investigating how to best make the switch.

Sensitivity analysis is typically the second step. The purpose is to obtain a "sense" of how much key variables (e.g., relative yields) can change before the system which is projected to be the best tillage/planting system changes. For example, for the 200-hectare farm following a corn-soybean rotation, the annualized economic advantage that would result from the adoption of the chisel plow conservation tillage system is estimated at $14.30/hectare--based upon our assessment of the evidence that yields are comparable, and there are no cost differences except for machinery and labor. Thus, corn and soybean yields could fall as much as 2.2 percent for conservation tillage relative to conventional tillage before the advantage would be eliminated.

If there is a clear increase in annual net return to land from the adoption of a conservation tillage/planting system relative to conventional tillage system, then the analysis must proceed to an evaluation of how and when the switch should occur. This is the third step.

Partial Replacement and Modification of Machinery Complement

The first step in a replacement analysis is to take a partial budgeting approach to assess possible changes. For example, for the 200-hectare farm with a corn-soybean rotation, the addition of the required chisel plow would

---

[8]The concept of "economically best" is developed by Rotz and Black in Chapter 16 of this volume.

result in an expenditure of $4,240. Using the annual cost factor (fraction of new cost/year) that is presented in Chapter 16, the resultant increase in annual use cost would be $848 per year, or $4.20 per hectare per year. The required modifications to the planter (add fluted coulters and associated frame in front of planter) would cost $1,200, with a resultant annual use cost of $240. That is $1.20 per hectare. That results in a total increase of $5.40 per hectare.

The increase in annual use cost of $5.40 per hectare associated with the purchase of the chisel plow and the planter modification must be compared with the resultant fuel and labor cost reductions. In this instance, the fuel cost decreases from $20.25 to $16.80 per hectare ($3.45) and the labor cost decreases from $22.65 to $20.69 per hectare ($1.96). Thus, the $5.41/ha fuel and labor savings are just equal to the additional machinery costs. If there are no reductions in yield per hectare and no pesticide or fertilizer cost increases under the chisel plow variant of conservation tillage for the Soil Management Groups being studied, then there is no reduction in projected net returns associated with the switch to the chisel plow machinery complement change that reduces off site damage.

Complete Replacement of Machinery Complement

When it can be shown, as a result of the comparative "all new equipment" analysis, that it is profitable to make an adjustment but not at the current time, then the farmer must examine the best time to switch from conventional to the chosen conservation tillage system. The objective is to maximize the net present value of future net cash flows (Perrin, 1972).[9] The switch rule is as follows (when there are no long-term reductions in potential productivity associated with a conventional moldboard plow system):

Step 1. Solve for the economically best trade-in age of the "challenging" conservation tillage/planting systems. That is, find the age at which the average annual cost of keeping the machinery complement in place is a minimum.

Step 2. Calculate the marginal cost of holding the existing "defender" conventional tillage system for

---

[9]Generally, analysts attempt to account for risk differences among systems. That is, if one system is more risky than the other, farmers typically require a higher rate of return before they are willing to switch.

an additional year. Then, compare the marginal cost of holding the defender with the average annual cost of the challenger. If it costs more to hold the existing conventional tillage system for an additional year than it does to replace it with the conservational tillage/planting system, the farmer should replace it. However, if the cost of holding the defending conventional tillage complement for an additional year is less than the average annual cost of the challenger, the farmer should continue to use the conventional system for at least an additional year.[10]

Extension

The results outlined can be extended to cover the case in which wind and water erosion reduce the intermediate and long-term potential productivity of the soil. Here, the annual cost of holding the defender for an additional year must include the added cost of the present value of the reduction in net return over the infinite future associated with the potential productivity decline that occurs. These were not considered in this study, since for the relatively level, fine textured soils, there is little evidence that erosion will reduce productivity.

ECONOMIC COMPARISON OF THE CHISEL PLOW VARIANT OF CONSERVATION TILLAGE VERSUS CONVENTIONAL TILLAGE

This section develops a synthesis of the information from the literature and from the farmer field comparison plots. This information, in conjunction with the machinery annual use cost developed in Chapter 16, is used to estimate the difference in the net returns to land which would be expected to result from the use of the chisel plow conservation tillage system as contrasted to the traditional conventional moldboard plow tillage system. The analysis is conducted on a pre-tax basis.

Synthesis of Information from the Farmer Comparison Plot and the Literature

Crop/Yield/Hectare. The literature suggested that on tilled, naturally poorly drained, fine-textured soils, corn yields for the chisel plow tillage system would range from no reduction to as much as a 10 percent reduction relative to yields associated with the fall moldboard plow tillage system. The extent of the reduction is influenced

---

[10]The execution of these steps is discussed in more detail in Black et al. (1984), Chapter 4.

by the drainage conditions, tilth of the soil, and cropping system. The results of the project area side-by-side field plot comparisons for corn suggest a very small yield reduction associated with the chisel plow system relative to the conventional tillage system, probably no greater than one percent, except in years that are unusually wet during the latter part of the growing season.

No yield reduction for the chisel plow tillage system, relative to the conventional system, is projected for navy beans. These results are consistent with both the side-by-side field comparison plots and replicated small plot experiments.

Sugar beet yields are assumed to be similar under both chisel plow and the conventional tillages. The number of observations from the comparison plots was too small to be meaningful, except to note that none suffered significant performance losses as a result of the adoption of conservation tillage.

Seeding Rate. The side-by-side field plot comparisons suggest final corn population is lower under conservation tillage, relative to conventional tillage. However, corn yields were nearly the same for the chisel plow and conventional tillage systems. Yields would be expected to increase slightly if seeding rates were increased. Our assumption is that if seeding rates increase, yields will increase and as a result, there will be no change in the "net" seed cost.

Pesticide and Herbicide Use. Projected to be the same for both tillage systems. There were essentially no differences in insect and weed problems between the chisel plow and the conventional tillage systems in the side-by-side comparison plots. The literature suggests there are increases sometimes, typically relatively small, associated with the use of the chisel plow tillage system.

Fertilizer Use. Projected to be the same for both tillage systems, particularly since we are assuming that anhydrous ammonia is knifed into the soil.

Labor Requirements. Projected labor requirements are less for the chisel plow tillage system than for the conventional tillage system. They are based upon the projections in Chapter 16. The assumptions about initial investment, planning horizon, discount rate, fuel costs, repairs, and housing were presented there.

Corn Drying Costs. Projected to be comparable for conservation tillage and conventional tillage under all but the wettest of harvest conditions. For purposes here, we are assuming no difference. Corn drying is projected to be $.98 for each percentage point of moisture removed per ton (Schwab et al., 1984).

Projected Commodity Prices. The commodity prices used in the analysis are based upon 10-year historical relationships. They are converted to 1984 prices, and are as follows:

Corn:	$104.10/ton
Wheat:	$137.13/ton
Navy Bean:	$548.02/ton
Sugar Beet:	$ 31.80/ton

The assumption in the projections is that it is difficult to predict what commodity prices will be in an individual year, but farmers in the project area make machinery complement choices and crop rotation decisions in the context of a longer-run target acreage mix. Thus, the relative prices among commodities that would be expected to hold the next decade, on the average, are the appropriate planning prices.

Economic Advantage of the Chisel Plow Tillage System

Partial Budgeting. Partial budgeting will be used in the comparative analysis. That is, only those cost and revenue items that change between systems will be considered. Based upon the synthesis presented in the previous section, these are the annual use and variable costs of machinery and labor.

New Cost Basis. If both the chisel plow and the conventional tillage systems were purchased new, there is a consistent and unambiguous increase in net returns for the chisel plow system (Tables 3 and 4). Stated alternatively, chisel plow system yields relative to conventional tillage systems, averaged across all crops, could fall between 2 and 4 percent before the cost reduction associated with the adoption of conservation tillage would be eliminated. If the yield reduction were for a particular crop such as corn in an corn-soybean or corn-navy bean rotation, the yield reduction could be as much as 4-5 percent before the cost savings would be eliminated.

The largest advantage to the chisel plow conservation tillage system occurs under the continuous corn cropping system. This is the result of the severe constraint on

Table 3. Cropping sequences for the Saginaw Bay Project.

Corn - Corn (for contrast only)
Corn - Navy Bean
Corn - Corn - Navy Bean
Corn - Soybean
Corn - Corn - Soybean
Corn - Navy Bean - Sugar Beet
Corn - Navy Bean - Navy Bean - Sugar Beet
Corn - Navy Bean - Wheat - Sugar Beet
Corn - Navy Bean - Soybean - Sugar Beet
Corn - Corn - Navy Bean - Wheat
Corn - Corn - Navy Bean - Sugar Beet
Oats - Alfalfa (clear seed, 1 yr stand) - Navy Bean - Sugar Beet
Alfalfa (clear seed, 3 yr stand) - Corn - Corn - Sugar Beet

Table 4. Average of machinery plus labor cost differences between conventional tillage and the chisel plow variant of conservation tillage for 200-500 hectare farms.*

Rotation	Advantage to the Chisel Plow Tillage System, $/ha	Break-even Reduction, Percent**
Continuous Corn	22.90	3.5
C-NB	11.60	1.8
C-C-NB	16.00	2.4
C-Soy	15.20	2.3
C-C-Soy	14.40	2.2
C-NB-SB	15.80	2.4
C-NB-NB-SB	13.50	2.1
C-NB-W-SB	12.40	1.9
C-NB-Soy-SB	14.40	2.2
C-C-NB-W	12.50	1.9
C-C-NB-SB	14.80	2.3
O-A-NB-SB	13.60	2.1
A-C-C-SB	21.40	3.3

*Farm size was evaluated at 50 ha increments

**The percentage yields, averaged across all crops, could be reduced before the advantage to conservation tillage would be eliminated.

time to complete operations that result from a single crop system. In general, the advantages to conservation tillage are highest when the system reduces bottlenecks in tillage operations.

Transition Basis. The previous section demonstrates that there is economic incentive to adopt the chisel plow tillage system. However, if adoption takes place in association with the existing machinery complement, new machinery and modifications must be purchased and there must be associated fuel and labor savings to justify the purchase. If the existing machinery complement were used, but a chisel plow was substituted for a moldboard plow and the planter were modified, there would be new annual use costs of $4.00-6.00 per hectare. The reduction of fuel and labor costs is approximately equal to this additional expense. Thus, there is not a large immediate economic incentive to adopt the chisel plow system, but the farmer's net return should be at least as large as it was before.

If the farmer already owns a chisel plow and is using it as a primary tillage tool following dry bean harvest, the side-by-side comparisons suggest the system should be used following corn harvest prior to planting beans. Also, the comparison data show that farmers in the project area can successfully use the chisel plow system on "heavier" soils than they previously believed was economically feasible.

If the addition of a chisel plow and modification of the planter is not economically feasible as a transition strategy, then the conventional replacement analysis would be followed. These calculations indicate that an existing machinery complement would need to be at least seven to ten years old before the farmer would expect to see a complete shift from conventional to the chisel plow conservation tillage system. The shift would take place sooner only for those farms in the process of growing.

SUMMARY

The purpose of this study was to evaluate the profitability of conservation, as contrasted to conventional, tillage in the Southeast Saginaw Bay Watershed. The task was divided into three components: (1) comparison of the performance on four-hectare side-by-side conventional and conservation tillage plots under practical management conditions on selected farms in the watershed; (2) synthesis of results from literature and field investigations; and (3) economic analysis from a whole-farm perspective based upon the synthesis.

The results of the field plot comparisons were broadly consistent with literature based upon smaller, replicated plot experiments. Corn yields under the chisel plow conservation tillage system averaged 1.5 percent less than yields under the moldboard plow conventional moldboard tillage system over the four years of the project. The differences were negligible except in 1981, an unusually wet year during which the soil was saturated during the latter part of the growing season. The yields of dry beans were comparable between the chisel plow and the conventional tillage systems.

Herbicide, insecticide, fertilization and seeding rates were maintained at the same level by the cooperating farmers. Weed and insect problem differences between chisel plow and conventional tillage systems were negligible for all crops.

There is a reduction in machinery plus labor costs if chisel plow conservation tillage is adopted--assuming the farmer is at a decision point, and can replace his existing conventional moldboard plow machinery complement. The economic advantage to conservation tillage is typically $15 to $20 per hectare. The size of the advantage is sensitive to crop sequence and farm size.

Economics favor the purchase of a chisel plow and modification of planter (for those that are suitable for modification) immediately--for those farms where no yield reductions, relative to conventional tillage, are projected and where no increases in other costs such as fertilizer, herbicides and insecticides are expected. The net economic gains, however, are small. To rephrase the point, chisel plow conservation tillage can be implemented immediately without reducing net farm income on most soil types in the watershed.

LITERATURE CITED

Black, R., D. Christenson, A. Rotz, H. Muhtar, and J. Posselius. 1984. Results of an economic comparison of conventional and conservation tillage systems in the Southeast Saginaw Bay Drainage Basin. Dept. Ag. Econ., Michigan State University, East Lansing, MI 48824.

Christenson, D., Z. Helsel, V. Meints, R. Black, R. Hoskins, F. Wolak, and T. Burkhardt. 1980. Agronomics and economics of some cropping systems for fine textured soils. Dry Bean Digest, Winter 1980.

Collins, R. and J.C. Headley. 1983. Optimal investment to reduce the decay rate of an income stream: The case of soil conservation. J. Envir. Econ. Manage. 10:60-71.

Dasgupta, P. 1982. The Control of Resources. Harvard University Press, Cambridge, MA 02138.

Doster, D., D. Griffith, J. Mannering, and S. Parsons. 1983. Econometric returns from alternative corn and soybean tillage systems in Indiana. J. Soil Water Conserv. 38:504-508.

Fisher, A.C. 1981. Resource and Environmental Economics. Cambridge University Press, New York, NY 10022.

Frohberg, K. and E. Swanson. 1979. A method for determining the optimum rate of soil erosion. Department of Agricultural Economics Report AERR 161, University of Illinois at Urbana-Champaign, April, 1979.

Griffith, D.R. and J.V. Mannering. 1985. Differences in crop yields as a function of tillage system, crop management and soil. In: F.M. D'Itri (ed.), A Systems Approach to Conservation Tillage, Lewis Publishers, Inc., Chelsea, MI 48118, pp. 47-59.

Muhtar, H. 1982. An economic comparison of conventional and conservation tillage systems in the Southeast Saginaw Bay Drainage Basin. Ph.D. Dissertation in Agricultural Engineering, Michigan State University, East Lansing, MI 48824.

Perin, R.K. 1972. Asset replacement principles. Amer. J. Agr. Econ. 54:60-67.

Pierce, F., W. Larson, R. Dowdy, and W. Graham. 1983. Productivity of soils: Assessing long-term changes due to erosion. J. Soil Water Conser. 38:39-44.

Rotz, C.A. and R. Black. 1985. Machinery requirements and cost comparisons across tillage systems. In: F.M. D'Itri (ed.), A Systems Approach to Conservation Tillage, Lewis Publishers, Inc., Chelsea, MI 48118, pp. 171-190.

Rotz, C.A., H. Muhtar, and R. Black. 1983. A multiple crop machinery selection algorithm. Transactions of the American Society of Agricultural Engineers 26(6):1644-1649.

Schwab, G. 1983. Custom work rates in Michigan. Michigan State University Cooperative Extension Service Bulletin No. 456, East Lansing, MI 48824.

Tilman, S. and D. Mokma. 1976. Soil management groups and soil erosion control. Agr. Exp. Sta. Research Report 310, Michigan State University, East Lansing, MI 48824.

Vitosh, M.L., W.H. Darlington, C. Rice, and D. Christenson. 1985. Fertilizer management for conservation tillage. In: F.M. D'Itri (ed.), A Systems Approach to Conservation Tillage, Lewis Publishers, Inc., Chelsea, MI 48118, pp. 89-98.

PART IV

CONSERVATION TILLAGE: ENVIRONMENTAL,
PUBLIC POLICY, AND SOCIOLOGICAL CONSIDERATIONS

CHAPTER 18

CONSERVATION TILLAGE: WATER QUALITY CONSIDERATIONS

James L. Baker
Department of Agricultural Engineering
Iowa State University
Ames, Iowa 50011

INTRODUCTION

Agricultural chemical losses from cropland with surface runoff, sediment, and subsurface flow represent environmental, economic, and energy concerns. Although chemcials are usually lost to some degree by each of these modes, depending on the chemical and soil properties, one mode usually dominates for a particular chemical. Therefore, chemicals can be grouped into three categories: those lost mainly with sediment, those lost mainly with surface runoff water, and those lost mainly by leaching.

While water quality considerations are usually expressed in terms of concentrations, losses or loads with agricultural drainage are important downstream because they affect the final concentrations in receiving waters. The chemicals of concern relative to nonpoint pollution from agricultural drainage generally involve the nutrients nitrogen and phosphorus, and pesticides. For surface water resources which must support aquatic life and serve as potentially potable water sources, there is concern for total nitrogen and phosphorus entering the system as well as for the specific ions NH_4^+ (NH_3, un-ionized NH_4^+, being toxic to fish), NO_3^- (conversion to NO_2^- causing methemoglobinemia in infants), and PO_4^{3-} (nutrient often limiting algal growth). In the case of groundwater resources that are being protected as sources of potable water, it is contamination with NO_3-N above 10 mg/l that is of the greatest concern. High levels of NH_4-N (greater than 0.5 mg/l) would also be of concern where chlorine is used as a disinfectant, because NH_4^+ reactions with chlorine result in compounds with much lower disinfecting efficiency than free chlorine.

With pesticides, there is also concern for both aquatic and human life. In general, herbicides are much less toxic than insecticides to both mammals and fish. The U.S. EPA (1976) has published concentration criteria on domestic water supply and freshwater and marine aquatic life for only two herbicides and 15 insecticides out of over 1,000 known pesticides. Domestic water supply concentrations for 2,4-D and 2,4,5-TP are set at 100 and 10 ug/l, respectively. Aquatic life criteria for most chlorinated hydrocarbon insecticides (aldrin, dieldrin, and DDT; all now banned) are extremely low, in the 0.001 ug/l range. Because of their persistence and potential carcinogenicity, it is recommended that human exposure to these insecticides be minimized. The organophosphorus insecticides listed (guthion, malathion, and parathion) are apparently one or two orders of magnitude less toxic to aquatic life; their criteria concentration range is from 0.01 to 0.1 ug/l.

Because there is a need to establish guidelines for additional pesticides possibly found in drinking water, the Hazard Evaluation Division of EPA's Office of Pesticide Programs has announced plans to establish maximum advisable levels (MAL) for pesticides in groundwater. The basis will likely be the toxicology data base and scientific expertise currently utilized to establish tolerances for pesticide residues in food. The concept of acceptable daily intake (ADI) would be extended for the assessment of potential hazard. The ADI is the daily exposure level of a pesticide residue that, taken during the lifetime of a man, appears to be without appreciable risk. It is generally expressed as mg of pesticide per kg of body weight per day. Under the EPA plan, the maximum advisable level in one liter of water would be set as equal to the ADI for the pesticide of interest times 10 kg. Using ADI values given by the National Academy of Sciences (1977), the MALs for heavily used pesticides such as atrazine, alachlor, trifluralin, carbaryl, and malathion would be 215, 1000, 1000, 820, and 200 ug/l, respectively; examples of pesticides with low MALs would be dicamba, aldicarb, and phorate with values of 12.5, 10 and 1 ug/l. Even small percentage losses of these latter pesticides in terms of that applied could result in concentrations exceeding these values. For instance, an annual loss with water of 0.2 percent of a 1 kg/ha application of phorate would contaminate drainage to a level of 1 ug/l if 20 cm of drainage occurred. Adsorption to or release from deposited or simultaneously transported sediments may change concentrations in final receiving waters.

Edge-of-field losses are determined by chemical concentrations times the volume or mass of the carrier, be it sediment or water. Thus, any management practice, such as conservation tillage, that affects the volume of the carriers, affects chemical losses. Erosion control, the main attribute of conservation tillage, reduces sediment volumes. For most variations of conservation tillage systems and for most soil and weather conditions, infiltration is increased and surface runoff water volumes are also reduced. However, as a result of increased infiltration, subsurface flow volumes can be expected to increase.

In addition to affecting the carrier volumes, conservation tillage affects chemical concentrations in sediment and water because of differences in hydrology, chemical placement, amount of chemical used, and the chemistry of the soil surface itself with residue left on it, relative to the conventional moldboard plow-disk-plant tillage system of the past. Table 1 presents a qualitative view of the effects of conservation tillage on chemical losses. As is shown, there are both benefits as well as potential

Table 1. Qualitative Water Quality Effects of Conservation Tillage

Conservation Tillage Effects	Chemicals lost with:		
	Sediment[1]	Runoff Water[1]	Leaching[1]
decreased erosion	-	o	o
more infiltration, less runoff	o	-	+
less incorporation of chemicals	+	+	?
higher surface organic matter, lower pH	+	?	?
surface crop residue	o	+	o
possible additional chemical use	+	+	+
Overall:	-	- to +	o to +

[1]+ indicates increased losses, - decreased losses, o no effect, and ? the effect could vary from + to -.

problems. Each of these effects is discussed in more detail in the following sections, with quantitative results given where possible. It is the purpose of this paper to show what water quality benefits are now being obtained with the use of conservation tillage, where potential water quality problems exist, and what might be done to alleviate them. The comparisons will be made with the moldboard plow system, which for purpose of discussion will be called the conventional system, even though it is no longer the most popular system in many areas.

DISCUSSION

Decreased Erosion

It has been shown that the percentage of the soil surface covered with residue and not the tillage tool used, if any, is the major factor determining soil loss (Laflen et al., 1978). An exponential equation has been developed to relate soil loss to percent residue cover (Laflen and Colvin, 1981). The current choice of coefficients for that equation would indicate that for every additional 20 percent of the soil surface covered by crop residue, erosion is decreased by half (Laflen, 1983). Therefore, if chemical concentrations in sediment were constant, chemical losses with sediment would also be cut in half. However, erosion is a selective process with the finer, more chemically active particles preferentially transported. Menzel (1980) has developed an equation from field data which shows that total nitrogen and phosphorus concentrations in sediment increased (termed enrichment) by a factor of 1.15 when erosion is decreased by half. As a result, if soil loss is decreased to 50 percent of its previous value, nitrogen and phosphorus losses with sediment would only be increased to 57 percent of their previous values.

Field data in Table 2 (taken from Baker and Johnson, 1983) illustrate the phenomena of enrichment. For example, in the first study listed, till-planting reduced soil loss to 38 percent of conventional tillage, but only reduced nitrogen loss with sediment to 54 percent of conventional tillage. For studies where total phosphorus losses were measured (as opposed to measurements of available phosphorus, discussed later under Less Incorporation of Chemicals), enrichment also was observed.

Because total nitrogen concentrations (range of about 1000 to 5000 ug/g) and total phosphorus concentrations (range of about 400 to 2000 ug/g) in sediment are much higher than nitrogen (usually < 10 mg/l) and phosphorus

Table 2. Runoff, erosion, and nutrient losses for conservation tillage expressed as a percentage of those for conventional tillage.

Practice	Study[1]	Soil	Slope %	Rnof %	Eros %	Nitrogen			Phosphorus			Ref.	Comment
						Sol'n %	Sed't %	Total %	Sol'n %	Sed't %	Total %		
Till-plant	N,W	sil	10-15	65	38	68	54	55	180	112[2]	130	(3)	cont. corn
No-till (ridge)				58	11	44	19	20	230	36	58		cont. corn
No-till	N,W	sil	9	9	1							(4)	cont. corn
No-till	N,P	sil	5	51	1	70	6	10	1400	6	16	(5)	beans-beans
No-till				38	1	190	6	21	1600	5	17		beans-wheat
No-till				106	12	180	40	52	450	36	39		beans-corn
No-till				80	3	410	14	47	1650	13	25		corn-beans
Till-plant	N,P	sicl	6	71	58							(6)	corn
Till-plant	S,P	sil	8-12	86	33	2100	41	92	2250	46	47	(7)	cont. corn (fert.treat.)

Table 2. Continued

Practice	Study[1]	Soil	Slope %	Rnof %	Eros %	Nitrogen Sol'n %	Nitrogen Sed't %	Nitrogen Total %	Phosphorus Sol'n %	Phosphorus Sed't %	Phosphorus Total %	Ref.	Comment
Chisel				49	5	1900	9	52	1950	10	11		cont. corn (fert.treat.)
Disk				85	15	1050	18	42	1850	16	17		cont. corn (fert.treat.)
No-till (coulter)				74	8	3950	10	99	100000	24	55		cont. corn (fert.treat.)
Till-plant	S,P	sil	5-12	83	77	200	68	70	315	165[2]	170	(8-9)	cont. corn
Chisel				96	62	205	59	62	335	130	135		cont. corn
Disk				84	31	215	38	41	390	84	93		cont. corn
No-till (ridge)				77	15	280	28	33	585	67	82		cont. corn
No-till (coulter)				75	8	270	14	19	625	32	50		cont. corn
Disk-chisel	S,P	sil	5	72	27	260	31	34	100	34	35	(10)	corn
Coulter-chisel				65	24	120	27	29	83	30	31		corn
Chisel				70	39	140	40	41	83	42	42		corn

Table 2. Continued

| Practice | Study[1] | Soil | Slope % | Rnof % | Eros % | ----Nitrogen---- | | | ----Phosphorus---- | | | Ref. | Comment |
						Sol'n %	Sed't %	Total %	Sol'n %	Sed't %	Total %		
Disk				70	20	240	26	29	75	26	27		corn
No-till				90	17	120	21	22	100	21	22		corn
Chisel-disk	S,P	sil	11	87	61							(11)	row-crop
No-till				109	36								row-crop
Chisel-disk	S,P	sl	5	69	45								row-crop
No-till				85	28								row-crop

[1]N indicates natural precipitation; S, simulated rainfall; W, watershed; and P, plot.
[2]Phosphorus lost with sediment was as available phosphorus, for other studies as total phosphorus.
[3]Johnson et al., 1979. [4]Harrold et al., 1970.
[5]McDowell and McGregor, 1980. [6]Onstad, 1972.
[7]Romkens et al., 1973. [8]Barisas et al., 1978.
[9]Laflen, et al., 1978. [10]Siemens and Oschwald, 1978.
[11]Laflen and Colvin, 1981.

(usually < 1 mg/l) concentrations in water, for runoff with sediment concentrations in runoff water in excess of about 10,000 mg/l, losses with sediment dominate total runoff losses. Therefore, total nitrogen and phosphorus would be categorized as chemicals lost mainly with sediment. As shown in Table 2, the erosion control provided by conservation tillage reduced total nitrogen loss from 1 to 90 percent and total phophorus losses from 45 to 89 percent over conventional tillage, with no-till generally being the most effective.

Data in Table 3 (taken from Baker and Johnson, 1983) show the effect of erosion control with conservation tillage on pesticide runoff losses. For each of the herbicides alachlor, propachlor, and atrazine, the ratio of concentrations in sediment to that in water in a runoff sample was usually in the range of about 4 to 10. Even with high erosion rates, most of the herbicide was lost with runoff water and not with sediment, so that erosion control did not result in herbicide runoff control. These herbicides would be considered moderately adsorbed by soil and would be categorized as lost mainly with surface runoff water. The insecticide, fonofos, is more strongly adsorbed to soil with sediment-water concentration ratios usually from 25 to 100, and in this case, erosion control resulted in reduced runoff losses except for the till-plant system. For both the natural and simulated rainfall studies, fonofos losses with till-plant exceeded those for conventional tillage despite decreased erosion and runoff. The insecticide was banded and incorporated in the row which is bare of residue in the till-plant system. It is probable that the majority of eroded soil came from this area, thus increasing fonofos losses. For the last study listed in Table 3, chemical enrichment in sediment (as sediment loss decreases) does not appear to be a factor for the moderately adsorbed herbicide alachlor, but does for the more strongly adsorbed fonofos. As a result, fonofos concentrations in runoff water for conservation tillage were also somewhat higher than those for conventional tillage. The slightly higher alachlor concentrations in runoff water for some conservation tillage systems are believed due to broadcast application to surface residue and will be discussed later.

More Infiltration, Less Runoff

The data in Table 2 show that conservation tillage usually reduces the volume of runoff by an average of about 25 percent, but the degree of reduction is highly variable. In four of the five studies considering both no-till (or coulter) and chisel treatments, no-till resulted

Table 3. Runoff, erosion, and pesticide losses for conservation tillage expressed as a percentage of those for conventional tillage.

Practice	Study[1]	Soil	Slope %	Rnof %	Eros %	Pest 1 — Sol'n %	Sed't %	Total %	Pest 2 — Sol'n %	Sed't %	Total %	Ref.	Comment
Till-plant	N,W	sil	10-15	65	38	47 (alachlor)	52	48	126 (fonofos)	110	116	(4)	cont. corn
No-till				58	11	30 (propachlor)	5	24	50 (atrazine)	12	26		cont. corn
No-till	N,W	sil	10-15	52	10	0[2] (simazine)	0	0	30 (atrazine)	5	25	(5)	cont. corn (contoured)
No-till	N,W	sil	6-20					4	(atrazine)		880[3]	(6)	cont. corn
Till-plant	S,P	sil	5-12	83	77	96 (alachlor)	77	91	115 (fonofos)	131	127	(7)	cont. corn
Chisel				96	62	150	62	125	163	64	90		cont. corn
Disk				84	31	115	31	92	127	49	70		cont. corn
Ridge				77	15	116	15	87	137	71	88		cont. corn
Coulter				75	8	127	8	93	145	19	52		cont. corn

[1] N indicates natural precipitation; S, simulated rainfall; W, watershed; and P, plot.
[2] No detectable losses from no-till.
[3] Very small losses were measured for conventional.
[4] Baker and Johnson, 1979. [5] Ritter et al., 1974. [6] Triplett et al., 1978. [7] Baker et al., 1978.

in slightly greater runoff volume. Even though the moisture content of the surface soil is usually greater for conservation tillage versus conventional tillage, the factors of preserving soil structure, including macropores with no-till, in conjunction with residue protecting the soil surface from sealing are believed to dominate and be responsible for more infiltration with conservation tillage. However, the effect of more infiltration, say on an annual basis, on surface runoff losses is more complicated than just the effect of the volume of carrier. Timing of runoff and its effect on concentrations is also a factor. For the first storm after tillage, the difference in runoff between no-till and any tillage, including the moldboard plow, is usually in favor of tillage, due to the void space and roughness created. In a study of two adjacent watersheds (Hamlett et al., 1984), the one with the most recent tillage had the least runoff except for very intense storms when runoff volumes were equal. In another study, runoff from a no-till area after soybean harvest averaged 17 mm for an 111 mm simulated rain, but averaged only 12 and 3 mm for disked and chiseled areas. Times to the beginning of runoff averaged 29, 46 and 71 minutes; and residue covers were 82, 31 and 48 percent, respectively, for the three tillage treatments. For a second rain of 72 mm twelve days later, runoff for no-till decreased to 13 mm, while runoff for the disked area increased to 34 mm and for the chiseled area to 9 mm (Baker and Laflen, 1983).

The first storm after chemical application usually results in the greatest surface runoff losses, and during a storm, concentrations of a chemical categorized as one lost mainly in surface runoff decrease with time as the chemical is removed (by both runoff and leaching) from the thin surface layer that interacts with rainfall and runoff (Wauchope, 1978; Baker, 1980). Therefore, the decreased total runoff in the first storm after tillage that is associated with a chemical application, as well as the increase in time before runoff begins, can decrease the loss of a surface-applied chemical relative to no-till. This decrease depends on the duration and intensity of the storm and should be greater for the tillage system that leaves more residue on the soil surface to protect against surface sealing. Whether the initially lower, but more constant infiltration rate for no-till results in greater or lower annual chemical losses depends on the number and magnitude of later runoff events. For a chemical lost mainly with sediment, concentrations in runoff are low as well as being more nearly constant with time so that changes in infiltration rates and totals do not significantly affect losses.

As with chemicals lost with surface runoff water, the effect of increased infiltration on chemicals leached is more complicated than just the effect of increased carrier volume. The routes and rates of infiltration are also factors. For example, to envision the differences between no-till and conventional tillage, consider two finely porous, ceramic columns, one representing conventional tillage, the other, which has small (2 mm) diameter holes drilled through its length, representing no-till. If the columns were wetted, a NO_3 salt added to the top, and the columns flushed with ponded distilled (or low NO_3) water, the "macropores" in the no-till column would allow the NO_3 to come through faster and in larger amounts. This was the case in one study (Tyler and Thomas, 1977) where NO_3-N losses in leachate under no-till corn were greater than for conventional corn. This was believed to be due to the surface-applied NO_3-N being washed into natural soil cracks and channels and flowing much deeper into the soil than predicted by miscible displacement theory. Likewise, decreased interaction of pesticides in water with soil adsorption sites at higher flow rates, similar to decreased interaction between water in micropores and water in macropores, resulted in increased downward movement of atrazine in one column study (Dao et al., 1979).

However, now consider the case where, rather than adding NO_3 in the form of a salt at the column surfaces, the two columns are wetted with an equal amount of NO_3 in solution, and then distilled water is ponded on the columns. In this example, the NO_3-N concentrations in outflow from the no-till column would be less. In this latter example, the effect of water bypassing much of the column holding NO_3 was used by Wild (1972) to partially explain the slow leaching of NO_3 originating from mineralization within aggregates in a fallow soil.

In a study of nutrient and atrazine losses with tile drainage for chisel versus moldboard plowing, Gold and Loudon (1982) found that the conservation tillage system had about the same total drainage but 21 percent more subsurface drainage and therefore less surface runoff than that for conventional tillage. Soluble phosphorus concentrations in tile drainage were greater for the conservation tillage system and were at a maximum for a major storm event occurring just four days after fertilization. Total phosphorus losses (including that with sediment) in total drainage was less for the conservation tillage system because of less soil erosion. Concentrations of NO_3-N in tile drainage were essentially equal for most events, but 1.6 kg/ha/yr more NO_3-N was lost in tile drainage from the conservation tillage area due mostly to the larger flow volume. Concentrations of surface-applied atrazine in tile drainage were also highest for the storm four days after application. For this storm, all drainage from the conservation tillage area was tile drainage. Over twice as much atrazine was lost with tile drainage from the conservation tillage area. Atrazine concentrations in surface runoff were 2 to 4 times higher than those in tile drainage.

Again, the timing and magnitude of storm events relative to the timing of chemical applications is important in determining the effect of conservation tillage, particularly no-tillage, on leaching losses of not only nitrate but phosphorus and pesticides. As evidenced by the previously discussed study, soaking rains immediately after surface application of chemicals can result in accelerated leaching due to movement through macropores with no-tillage or increased infiltration with increased residue cover for other conservation tillage systems. On the other hand, one or several small rains could occur first that would allow the chemical to enter the soil and diffuse through aggregates, and be less available for solution and leaching at a later time. One point to keep in mind in considering the effect of conservation tillage on leaching is that only about the top 20 cm layer of the total soil profile is affected by tillage.

Less Incorporation of Chemicals

Soil incorporation of chemicals by tillage is reduced or eliminated with conservation tillage because of the desire to leave crop residue on the soil surface. Concentrations and losses of soluble nutrients in runoff from areas where fertilizer has been incorporated by moldboard plowing have been found to be the same as for unfertilized areas (Timmons et al., 1973). This is also sometimes true for fertilizers incorporated by chisel plowing or disking (Baker and Laflen, 1983), but not always (Mickelson et al., 1983). Several other studies (Baker and Laflen, 1982; McDowell and McGregor, 1980; Johnson et al., 1979; Barisas et al., 1978; and Romkens et al., 1973) have found that soluble nutrient losses with conservation tillage, particularly no-till, are greater than those for conventional tillage because of decreased fertilizer incorporation. The stratification that occurs with reduced or no incorporation of applied nutrients results in increased nutrient concentrations in sediment as well as in water. Whether losses increase with sediment depends on the relative concentration increase and erosion decrease with conservation tillage.

In a study where phosphorus fertilizer was band-incorporated rather than surface-broadcasted (Mueller et al., 1982), soluble phosphorus concentrations were similar across all tillage systems, including conventional and no-till, and surface runoff losses from areas with conservation tillage were reduced in proportion to runoff volume reductions.

Lack of pesticide incorporation has also been found to result in higher concentrations and losses in both sediment and water. In one study (Bovey et al., 1978), picloram concentrations in surface runoff were about 3.6 times higher when the herbicide had been surface applied relative to a subsurface application. In another study (Baker and Laflen, 1979), atrazine, alachlor, and propachlor losses in runoff from an area with an unincorporated herbicide application were about three times higher than those for an area where herbicides were disk incorporated. Losses were even higher when herbicides were surface applied without incorporation on areas that had tractor tracks and surface compaction.

There is a tendency to assume that chemical incorporation would increase leaching because it shortens the path length to the water table; however, the effects of interaction of water in micropores and soil adsorption discussed in the preceding section must be considered in each case for incorporated versus surface applications. There is at least one ongoing study (at Iowa State University) to evaluate this issue.

Higher Surface Organic Matter, Lower pH

As reported by Dick (1983) and Blevins et al. (1983), lack of incorporation of crop residue and fertilizer with no-till or decreased incorporation with other forms of conservation tillage relative to moldboard plowing, results in higher concentrations of organic carbon, nitrogen and phosphorus and lower pH in the surface soil. For example, in the study by Dick, the surface of the Wooster silt loam soil after 19 years in no-till had organic carbon, nitrogen and phosphorus concentrations about 2.4, 2.2, and 1.2 times, respectively, those for conventional tillage. The pH decrease believed associated with nitrification of surface-applied nitrogen fertilizer was about 0.1 unit. Blevins et al. (1983) found that after 10 years of continuous corn production under no-till, the surface of a Maury silt loam soil had about twice the organic matter and pH 0.7 units lower than that for conventional tillage.

It is logical that the higher nutrient content of the surface soil for conservation tillage would result in higher nutrient concentrations in sediment because the surface soil is the major source of eroded sediment. For surface-applied pesticides, the higher organic matter content would generally result in increased adsorption and, therefore, higher concentrations and losses with sediment. However, as discussed by Baker and Johnson (1983), the effect of increased adsorption on losses with runoff water could be in either direction depending on where in the adsorption range the pesticide fell. For basic pesticides, such as atrazine, the lower pH with conservation tillage could also result in stronger adsorption of the positively charged ions to the negative clay surfaces and increased losses with sediment.

Another aspect of conservation tillage that could affect persistence of pesticides is the increased microbiological activity in the surface soil (Doran, 1980). Decreased persistence due to increased micobiological degradation could increase recommended rates of application to achieve pest control. As yet, no company is recommending increased pesticide application rates because of conservation tillage and this logic.

Surface Crop Residue

Crop residue on the soil surface can itself have two effects on chemical losses. First, it can be a source of nutrients to be released to surface runoff water washing over and running through it. Second, by it presence on the soil surface, it can act as a physical barrier preventing applied chemicals from reaching the soil surface, particularly broadcast sprayed herbicides.

Several studies, including those by Timmons et al. (1970) and Tukey and Romberger (1959), have shown that both living and dead plant tissues can release nutrients to water leaching through them, and McDowell and McGregor (1980) ascribed at least a portion of the increased soluble nutrient losses with no-till to release of nutrients from soybean residue. However, in studies where surface residue and surface fertilization were both factors (Baker and Laflen, 1982 and 1983), the impact of surface fertilization dominated runoff water quality relative to nitrogen and phosphorus. In one Iowa watershed study (Alberts et al., 1978), water and sediment weighted nutrient concentrations were highest during the December through March period when snowmelt occurred and there was corn residue on the soil surface; however, the values at this time were only 20 to 60 percent higher than annual averages. In another

Iowa watershed study (Baker et al., 1978), soluble nutrient concentrations in snowmelt runoff from watersheds with corn and soybean residue on the soil surface were quite similar to annual averages, including rainfall runoff after tillage incorporation of crop residues. As with applied chemicals, the timing of the release of any nutrients from crop residue relative to runoff-producing events (either rainfall or snowmelt), and the amount of water that infiltrates into the soil before runoff begins will be important in determining if losses are increased because of surface residue.

Herbicides applied to crop residue can be removed by either volatilization or washoff. Baker et al. (1978) found that concentrations in runoff from a simulated rain shortly after herbicide application increased as the percentage of the soil surface covered with corn residue increased. They believed washoff directly from the residue was responsible. Martin et al. (1978), found that there was little interaction between the four herbicides they studied and corn residue, with about 1/3 of the applied herbicide washing off with 5 mm of water and about 3/4 washing off with 35 mm of water. In a watershed study under natural rainfall conditions (Baker and Johnson, 1979), the occurrence of small rains after herbicide application but before a runoff-producing rain, apparently washed the herbicide from the corn residue to the soil with the conservation tillage systems, resulting in no effect of the residue on concentrations in runoff. The concern for herbicide volatilization from the crop residue has been expressed (Martin et al., 1978; Baker and Laflen, 1983), and work is under way at Iowa State University to quantify it.

Concentrations of Chemicals Dissolved in Drainage Relative to Standards

Ammonium. Ammonium concentrations in subsurface flow from cropland are usually low (<0.2 mg/l, as summarized by Baker, 1980) and below the 2 mg/l likely to produce enought ammonia to cause fish toxicity problems. Concentrations in surface runoff, although higher than in subsurface flow, are usually less than concentrations in rainwater producing runoff (often about 1 mg/l), because of extraction by the soil (Baker, 1980). This may not be true for runoff events occurring shortly after surface application with little or no incorporation of urea- or ammonium-based fertilizers. This is not a problem for conventional tillage where plow-down is used, but is a particular problem for no-till. In one rainfall simulation study (Baker and Laflen, 1983), the average ammonium concentration in runoff from a no-till area treated six days earlier with 31 kg/ha ammonium averaged 3.1, whereas the concentration for a fertilized and chisel plowed area was 0.5 mg/l, the same as for an unfertilized plot. In another study under similar conditions, but where soil antecedent moisture content was higher, resulting in less infiltration and more runoff sooner (Mickelson et al., 1983), ammonium concentrations for a surface-fertilized, no-till area, and a chisel-plowed, fertilized area were much higher at 36.0 and 2.8 mg/l, respectively. A check area had an average concentration of 0.6 mg/l.

Nitrate. Nitrate concentrations in surface runoff are usually below the 10 mg/l standard for drinking water, but concentrations in subsurface drainage from row-cropped land are usually at or above 10 mg/l (Baker, 1980). The effect of recent surface nitrogen fertilization can result in surface runoff concentrations exceeding 10 mg/l where initial infiltration is low due to tight or wet soils. In the study by Mickelson et al. (1983) discussed above, nitrate concentrations averaged 27.1, 3.7 and 0.3 mg/l for a surface-fertilized, no-till area, a chisel-plowed, fertilized area, and a chisel-plowed, check area, respectively.

Phosphate. Phosphate concentrations in surface runoff from fertile soils commonly exceed the 0.01 to 0.05 mg/l necessary to promote the growth of algae, whereas subsurface drainage usually has lower values than surface runoff, sometimes below 0.05 mg/l because of phosphate extraction from water by subsoils. Similar to ammonium and nitrate, recent surface phosphorus fertilization with little or no soil incorporation can substantially increase phosphate to concentrations in excess of 1 mg/l. However, unlike nitrogen, phosphorus fertility of the surface soil can build up resulting in sustained high concentrations in surface runoff over a season or longer as opposed to a few days or weeks for nitrogen.

Pesticides. Pesticide concentrations in surface runoff are highly variable with respect to group and time. Because pesticides do not occur naturally and do degrade with time after application (usually within 1 year for those currently used), pesticide concentrations in the first runoff event after application usually represent maximum concentrations. For those pesticides in the group lost mostly in surface-runoff water (medium adsorption class) such as atrazine, alachlor, cyanazine, metribuzin, propachlor, fonofos, and carbofuran, maximum concentrations were usually less than 1000 ug/l (Wauchlope, 1978), with flow-weighted concentrations on a seasonal basis much less (Baker, 1980). As an example, in four studies involving atrazine (Ritter et al., 1974; Baker and Johnson, 1979;

Burwell et al., 1974; Hall et al., 1972) average concentrations in runoff ranged from 147 to 0 ug/l.

For those pesticides in the group likely to be lost mostly in subsurface flow (least adsorption class) such as 2,4-D and dicamba, maximum concentrations in runoff were in the 2000-5000 ug/l range when applied mainly to foliage but were three times less when applied to bare soil. Maximum concentrations for those pesticides lost mainly with sediment (strongest adsorption class), such as trifluralin, were less than 25 ug/l.

Measurements of pesticides in subsurface drainage water (Muir and Baker, 1976; Baker and Austin, 1982; Hallberg et al., 1983) usually show lower concentrations than in surface runoff water and are in the range of 0 to 50 ug/l for all groups of pesticides. Gold and Loudon (1982) also found atrazine concentrations in surface runoff to be higher than in subsurface drainage, but their concentration range for subsurface drainage was 80 to 170 ug/l.

Whether a pesticide will likely exceed a standard concentration is highly dependent on the properties of the pesticide and on site conditions. As standards are established, decisions as to what alternative management practices are necessary to prevent problems can be made. However, if the 215 and 1000 ug/l values are established as MALs for atrazine and alachlor, respectively, no change in management may be necessary because losses and concentrations of these herbicides are already low enough. However, for compounds such as dicamba and phorate with low MALs, alternative practices such as decreased rates, mandatory soil incorporation or even restricted use in some areas may be necessary to reduce concentrations in drainage waters to acceptable levels.

SUMMARY AND CONCLUSIONS

The water quality advantages of reduced erosion with conservation tillage, and therefore reduced sediment and sediment-associated chemical losses, are significant and obvious. Reduced runoff that often occurs with conservation tillage also has the potential to decrease chemical losses with surface runoff water. However, reduced soil incorporation by tillage of surface-applied chemicals in the desire to leave crop residue on the soil surface increases chemical concentrations in runoff and potentially increases losses (depending on the relative magnitudes of increased concentrations versus decreased runoff volumes).

The solution to this potential problem is development of methods to incorporate chemicals without incorporation of residue. For example, Solie et al. (1981) discuss a blade that allows subsurface injection of herbicides without soil inversion and residue incorporation. This would also alleviate the concern for volatilization of herbicides surface-applied to crop residue. Dawelbeit et al. (1981) discuss a rolling, spoked-wheel, point-injector fertilizer applicator that injects liquid fertilizer beneath the soil surface, again without residue incorporation. There are also commercially available coulter and knife applicator assemblies for applying granular and liquid fertilizer in bands beneath the soil surface with little disturbance of surface residue.

The potential problem with possible additional chemical use with conservation tillage is probably overstated. For nitrogen, incorporation by the means discussed above will eliminate tie-up by microorganisms at the surface and, therefore, the need to apply additional nitrogen. Likewise, incorporation of potassium will eliminate the stratification at the surface. Although replacement of tillage for weed control early in the spring with the use of additional herbicide may increase runoff losses, a combination of banding and cultivation could decrease later herbicide use.

Increased infiltration with conservation tillage is a water quality benefit for those chemicals transported mainly with sediment and with runoff water (chemicals strongly or moderately adsorbed). The concern for nitrate leaching with increased infiltration could be decreased with improved nitrogen management (which is already needed even for conventional tillage areas). A better match between nitrogen availability and plant needs could be obtained by better timing, including multiple rather than a single annual application, possibly lower total rates, and use of formulations or additives that retain nitrogen in the ammonium form. Use of slightly adsorbed pesticides with low MALs may have to be banned in areas where the leaching potential is high due to excess precipitation and/or light, low organic matter soils.

Overall, it would appear that the soil conservation and potential water quality benefits of conservation tillage far outweigh potential problems--problems that should be solved with developing technology. Given the great inertia in shifting to acceptance of conservation tillage over the moldboard plow tillage system, it would seem wise to be watchful for currently unforeseen problems, but to not overemphasize unsubstantiated potential problems.

LITERATURE CITED

Alberts, E.E., G.E. Shuman, and R.E. Burwell. 1978. Seasonal runoff losses of nitrogen and phosphorus from Missouri Valley loess watersheds. J. Envirn. Qual. 7: 203-208.

Baker, J.L. 1980. Agricultural areas as nonpoint sources of pollution. Pages 275-310 in M.R. Overcash and J.M Davidson, eds. Environmental Impact of Nonpoint Sources Pollution. Ann Arbor Sci. Publ., Ann Arbor, MI.

Baker, J.L., and H.P. Johnson. 1983. Evaluating the effectiveness of BMP's from field studies. Pages 281-304 in R.W. Schaller and G.W. Bailey, eds. Agricultural Management and Water Quality. Iowa State Univ. Press, Ames.

Baker, J.L. and H.P. Johnson. 1979. The effect of tillage systems on pesticides in runoff from small watersheds. Trans., ASAE 22:554-559.

Baker, J.L., H.P. Johnson, M.A. Borcherding, and W.R. Payne. 1978. Nutrient and pesticide movement from field to stream: A field study. Proc., 1978 Cornell Agric. Waste Management Conf. on Best Management Practices for Agric. and Silviculture. pp. 213-245.

Baker, J.L., and J.M. Laflen. 1982. Effects of corn residue and fertilizer management on soluble nutrient runoff losses. Trans., ASAE 25:344-348.

Baker, J.L., and J.M. Laflen. 1983. Water quality consequences of conservation tillage. J. Soil and Water Cons. 38:186-193.

Baker, J.L., and J.M. Laflen. 1983. Runoff losses of nutrients and soil from ground fall-fertilized after soybean harvest. Trans., ASAE 26:1122-1127.

Baker, J.L., and J.M. Laflen. 1979. Runoff losses of surface-applied herbicides as affected by wheel tracks and incorporation. J. Environ. Qual. 8:602-607.

Baker, J.L., J.M. Laflen, and H.P Johnson. 1978. Effect of tillage systems on runoff losses of pesticides: a rainfall simulation study. Trans. ASAE 21:886-892.

Baker, J.L., and T.A. Austin. 1982. Impact of agricultural drainage wells on groundwater quality. Annual report, EPA grant no. G007228010. Iowa State Univ., Ames 50011.

Barisas, S.G., J.L. Baker, H.P. Johnson, and J.M. Laflen. 1978. Effect of tillage systems on runoff losses of nutrients: a rainfall simulation study. Trans., ASAE 21:893-897.

Blevins, R.L., M.S. Smith, G.W. Thomas, and W.W. Frye. 1983. Influence of conservation tillage on soil properties. J. Soil Water Conserv. 38:301-305.

Bovey, R.W., C. Richardson, E. Burnett, M.G. Merlke, and R.E. Meyer. 1978. Loss of spray and pelleted picloram in surface runoff water. J. Environ. Qual. 7:178-180.

Burwell, R.E., G.E. Schuman, R.F. Piest, R.G.Spomer, and T.M. McCalla. 1974. Quality of water discharged from two agricultural watersheds in Southwestern Iowa. Water Resour. Res. 10:359-365.

Dao, T.H., T.L. Lavy, and R.C. Sorensen. 1979. Atrazine degradation and residue distribution in soil. Soil Sci. Soc. Am. J. 43:1129-1134.

Dawelbeit, M., J.L. Baker, and S.J. Marley. 1981. Design and development of a point-injector for liquid fertilizer. Paper 81-1010. Am. Soc. Agr. Eng., St. Joseph, MI 49085.

Dick, W.A. 1983. Organic carbon, nitrogen, and phosphorus concentrations and pH in soil profiles as affected by tillage density. Soil Sci. Soc. Am. J. 47:102-107.

Doran, J.W. 1980. Soil microbial and biochemical changes associated with reduced tillage. Soil Sci. Soc. Am. J. 44:765-771.

Environmental Protection Agency. 1976. Quality criteria for water. Washington, D.C.

Gold, A.J. and T. L. Loudon. 1982. Nutrient, sediment, and herbicide losses in tile drainage under conservation and conventional tillage. Paper No. 82-2549, ASAE, St. Joseph, MI 49085.

Hall, J.K., M. Paulus, and E.R. Higgins. 1972. Losses of atrazine in runoff water and soil sediment. J. Environ. Qual. 1:172-176.

Hallberg, G.R., B.E. Hoyer, E.A. Bettis, and R.D. Libra. 1983. Hydrogeology, water quality, and land management in the Big Spring Basin, Clayton County, Iowa. Annual report, DEQ Contract No. 85-5500-02, Iowa Geological Survey, Iowa City 52242.

Hamlett, J.M., J.L. Baker, S.C. Kimes, and H.P. Johnson. 1984. Runoff and sediment transport within and from small agricultural watersheds. Trans., ASAE 27:1355-1363,1369).

Harrold, L.L., G.B. Triplett, Jr., and W.M. Edwards. 1970. No-tillage corn-characteristics of the system. Agr. Eng. 51:128-131.

Johnson, H.P., J.L. Baker, W.D. Shrader, and J.M. Laflen. 1979. Tillage system effects on sediment and nutrients in runoff from small watersheds. Trans., ASAE 22:1, 110-1, 114.

Laflen, J.M., J.L. Baker, R.O. Hartwig, W.F. Buchele, and H.P. Johnson. 1978. Soil and water loss from conservation tillage systems. Trans., ASAE 21:881-885.

Laflen, J.M. and T.S. Colvin. 1981. Effect of crop residue on soil loss from continuous row cropping. Trans., ASAE 24:605-609.

Laflen, J.M. 1983. USDA Agricultural Engineer, Iowa State University, Ames, IA 50011. Personal communication.

Martin, C.D., J.L. Baker, D.C. Erbach, and H.P. Johnson. 1978. Washoff of herbicides applied to corn residue. Trans., ASAE 21:1,164-1,168.

McDowell, L.L. and K.C. McGregor. 1980. Nitrogen and phosphorus losses in runoff from no-till soybeans. Trans., ASAE 23:643-648.

Menzel, R.G. 1980. Enrichment-ratios for water quality modeling. In W.G. Knisel (ed.) CREAMS, a field scale model for chemicals, runoff, and erosion from agricultural management systems. Vol. III, Rpt. No. 26. U.S. Dept. Agr., Washington, DC.

Mickelson, S.K., J.L. Baker, and J.M. Laflen. 1983. Managing corn residue to control soil and nutrient losses. Paper 83-2161. Am. Soc. Agr. Eng., St. Joseph, MI 49085.

Mueller, D.H., T.C. Daniel, B. Lowery, and B. Andraski. 1982. The effect of conservation tillage on the quality of the runoff water. Paper No. 82-2022, ASAE, St. Joseph, MI 49085.

Muir, D.C. and B.E. Baker. 1976. Detection of triazine herbicides and their degradation products in tile-drain water from fields under intensive corn (maize) production. J. Agric. Food Chem. 24:122-125.

National Academy of Sciences. 1977. Drinking water and health. Printing and Publishing Office, NAS, 2101 Constitution Avenue, Washington, DC.

Onstad, C.A. 1972. Soil and water losses as affected by tillage practices. Trans., ASAE 15:287-289.

Ritter, W.F., H.P. Johnson, W.G. Lovely, and M. Molnau. 1974. Atrazine, propachlor, and diazinon residues on small agricultural watersheds. Environ. Sci. Tech. 8:38-42.

Romkens, M.J.M., D.W. Nelson, and J.V. Mannering. 1973. Nitrogen and phosphorus composition of surface runoff as affected by tillage method. J. Environ. Qual. 2:292-295.

Siemens, J.C. and W.R. Oschwald. 1978. Corn-soybeans tillage systems: Erosion control, effects on crop production cost. Trans., ASAE 21:293-302.

Solie, J.B., H.D. Wittmus, and O.C. Burnside. 1981. Subsurface injection of herbicides for weed control. Paper 81-1011. Am. Soc. Agr. Eng., St. Joseph, MI 49085.

Timmons, D.R., R.E. Burwell, and R.F. Holt. 1973. Nitrogen and phosphorus losses in surface runoff from agricultural land as influenced by placement of broadcast fertilizer. Water Resources Res. 9:658-667.

Timmons, D.R., R.R. Holt, and J.J. Latterell. 1970. Leaching of crop residues as a source of nutrients in surface runoff water. Water Resources Res. 6:367-375.

Triplett, G.B., B.J. Conner, and W.M. Edwards. 1978. Herbicide runoff from conventional and no-tillage cornfields. Ohio Rep. Res. Devel. 63:70-73.

Tukey, H.B., Jr. and J.A. Romberger. 1959. The nature of substances leached from foliage. Plant Physiol. 34:vi.

Tyler, D.O. and G.E. Thomas. 1977. Lysimeter measurements of nitrate and chloride losses from soil under conventional and no-tillage corn. J. Environ. Qual. 6:63-66.

Wauchope, R.D. 1978. The pesticide content of surface water draining from agricultural fields--a review. J. Environ. Qual. 7:459-472.

Wild, A. 1972. Nitrate leaching under bare fallow at a site in northern Nigeria. J. Soil Sci. 23:315-324.

CHAPTER 19

ENVIRONMENTAL IMPLICATIONS OF CONSERVATION TILLAGE: A SYSTEMS APPROACH

George W. Bailey
Lee A. Mulkey
Robert R. Swank, Jr.
Environmental Research Laboratory
U.S. Environmental Protection Agency
Athens, Georgia 30613

ABSTRACT

Conservation tillage is projected to be the major soil protection method and candidate best management practice for improving surface water quality. Environmental and health implications as well as the agronomic virtues of conservation tillage must be identified and evaluated. A conceptual framework--mass balance approach--is developed identifying those system variables influenced by conservation tillage. A qualitative assessment is then made of the impact, mainly off-site in nature, of conservation tillage on the various exposure pathways of pesticides to human and aquatic ecosystems. Results from such an analysis suggest an increased potential for atmospheric losses of pesticides, a decrease in runoff and soil-erosion-related losses, but an increase in plant-residue-associated erosion as well as increased potential leaching through the unsaturated zone to groundwater. Major concerns, therefore, focus on the quantitative tradeoffs between runoff losses in water, leaching losses to groundwater, and enhanced atmospheric releases. A literature review, in general, substantiated the above speculations as to the impact of conservation tillage on soil properties, processes and activities, and on pesticide transformation and transport.

Possible environmental implications of conservation tillage that are mainly off-site in nature include: (1)

increased direct exposure to farm workers and rural residents due to enhanced volatilization; (2) increased threat of groundwater pollution due to enhanced infiltration and leaching of soluble, persistent pesticides; (3) enhanced instream impact on sensitive ecosystems due to higher pesticide concentrations in runoff and bound on clay-sized particulates; (4) decreased stream bank and stream bed stability; and (5) enhanced algal bloom tendency due to reduced total sediment loads and increased light penetration.

INTRODUCTION

In the last decade, the relationship between agricultural operations and water quality and quantity has become a major national concern. This concern is reflected in efforts to apply effective management practices to control, reduce or alleviate agriculture's contributions to nonpoint source pollution.

Pollutant discharges from agricultural sources are diffuse in nature and occur primarily during rainfall events. In the resulting runoff or leachate are found pesticides, nutrients, sediments, pathogens and easily oxidizable organics that may adversely affect water supplies, either from a human or from an environmental perspective.

To reduce the amount of pollutants entering streams, air or groundwater, a number of control methodologies, known as Best Management Practices (BMPs), were identified in the Federal Water Pollution Control Acts of 1972 and 1977. These BMPs provide a means of controlling the agronomic aspect of the agro-ecosystem that is the source of pollution and to a lesser extent the hydrologic and aerodynamic properties of the soil-plant system that regulates the rate of delivery and quantity of the pollutants reaching surface water, groundwater and the atmosphere.

A wide array of management practices have been developed to maximize agricultural productivity. These practices can be divided into two classes: (1) those that affect the physical system (e.g., tillage and structural controls) by influencing erosion through changes in soil erodibility and cover, slope steepness and slope length, and by influencing runoff through alterations in the soil infiltration rate, moisture storage capacity, and hydraulic conductivity, and (2) those that affect the properties of nonconservative pollutants or control their input relative to timing and level of application (e.g., environmental properties, formulations and application methodologies). Both conventional and conservation tillage fall into the first class. Conservation tillage is actually an array of practices that decreases the amount of exposed soil surface by retaining crop residues and reducing soil disturbance.

Research by Unger (1977, 1978, 1979), Crosson (1981) and Crosson and Brubaker (1982) projected that conservation tillage will be used on 50 to 60 percent of the tillable cropland in the United States by the end of the century. Estimates by the U.S. Department of Agriculture have run as high as 95 percent. The adoption of this "single management practice," therefore, would affect 85 to 162 million hectares of row crops. Agriculturally, this would be a quantum change in management systems comparable to that which occurred in the 1950s, when traditional crop-rotation-based farming was transformed into synthetic-chemical-based agriculture.

Such an extensive change in agricultural management gave rise to concerns about the total environmental impact of candidate BMPs (Bailey and Waddell, 1979; Bailey and Swank, 1983). Conservation tillage could offer large potential reductions in sediment loss to surface waters, but also could result in increased pesticide exposure to the environment through runoff to streams and impoundments, percolation to groundwater and loss to the atmosphere.

One of the indicated purposes of this conference is to define and evaluate the impact of the nearly ubiquitous implementation of conservation tillage on both environmental quality and human health.

The objectives of this paper are to:

(1)   develop a conceptual, mass-balance framework to assess the potential impact of conservation tillage from a pesticide perspective on the environment and human health,

(2)   make a qualitative analysis of the changes in pesticide exposure routes due to use of conservation tillage, and

(3)   elucidate the environmental implications of the widespread implementation and use of conservation tillage.

AGRO-ECOSYSTEM DESCRIPTION

Understanding the relationship between agricultural management practices and environmental quality requires an analysis of the complete agro-ecosystem. System analysis permits an enhanced understanding of the many dynamic

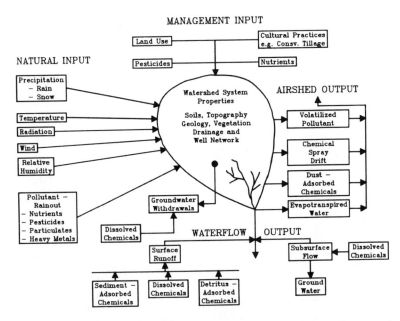

Figure 1. Factors influencing the behavior and export of agricultural chemicals from an agricultural watershed (after Bailey and Swank, 1983). (Reproduced from Agricultural Management and Water Quality, F.W. Schaller and G.W. Bailey, Editors, 1983, page 34, by permission of Iowa State University Press, Ames, Iowa.)

interactions among agro-ecosystem components from a physico-chemical standpoint. This approach is holistic; that is, an isolated study of the parts will not permit the comprehensive understanding of the complete system because the separate parts are linked interactively. A perturbation, e.g., widespread change in management practices in one part of the system, will not only affect that part but the whole system--terrestrial, aquatic and atmospheric compartments.

An agro-watershed has definite physical boundaries, and its response to pesticides or other chemical pollutants is determined by the combination of physico-chemical charac-teristics of the soils and chemicals themselves, topography, vegetation, geology and well and drainage networks. Inputs can be categorized as deterministic (e.g., timing of pesticide application) or stochastic (e.g., occurrence of rainfall). The system response, therefore, is stochastic. Outputs include food and fiber plus sediment, pesticides, nutrients, heavy metals and easily oxidized organics, plus residues on the land surface, in runoff and leachate and in the air.

The agro-ecosystem is a combination of hydrologic, agronomic, ecological, social and economic subsystems. The watershed level response, as seen from Figure 1, can be multimedia in dimension (land to water, land to atmosphere and atmosphere to land).

The hydrological cycle plays an important role in the transport of pesticides through a riverine system to its estuary. Possible pesticide transport routes include surface runoff, interflow to surface water, percolation to groundwater and direct loss to the atmosphere.

Chemical characteristics of the pesticide that influence pollutant transport from soil include: (1) ionic state (cationic, anionic, basic or acidic), (2) water solubility, (3) vapor pressure, (4) partition coefficient, (5) hydrophobic/hydrophilic character and (6) chemical and biological reactivity. These properties combined with the initial amount of pesticide applied determine how much pesticide potentially can be transported along each route. Sorption primarily determines whether the primary transport route is through overland flow, subsurface water movement, release to the atmosphere or sediment-bound movement (both water and aeolian modes of transport).

FRAMEWORK TO ASSESS POTENTIAL IMPACT OF CONSERVATION TILLAGE ON THE ENVIRONMENT AND HUMAN HEALTH

A systematic analysis of the environmental impact of conservation tillage must be based on defined endpoints for impacts and on a step-by-step approach showing how each endpoint is reached. The basic premise of our approach is the need to know the environmental and human health risks derived from widespread adoption of conservation tillage. The risk posed to any species is rather simply defined as the probability of damage, death, illness, functional impairment, ecosystem alterations, etc. Such damage only occurs when the affected specie is exposed to a harmful agent, usually in a probabilistic sense.

The major concern for health and environmental damage from conservation tillage is the associated use of pesticides. In short, how does conservation tillage influence the probability of pesticide exposure to biological species (including man) and what is the overall ecosystem function/response? The routes or pathways of exposure setting

Table 1.   Environmental pathways for chemical contaminants.

Pesticide Source	Transport Process	Exposure Pathway	Receptor (Population at risk)
canopy	volatilization	inhalation/ skin contact	human animals
crop residue	volatilization	inhalation/ skin contact	human animals
soil surface	volatilization	inhalation/ skin contact	human animals
root zone and below	volatilization	inhalation/ skin contact	human animals
grain/foliage	manufacturing/ feeding*	food/ingestion	human animals
crop residue	overland flow*	surface water (potable water)/ ingestion	human fish
runoff	overland flow*	surface water (potable water and food chain)/ ingestion	human fish
eroded soil particles	overland flow	surface water (food chain)/ ingestion, contact	fish
leachate	leaching/ percolation*	groundwater (potable water)/ ingestion	human animals

*Environmental flux of concern to either on-site or off-site receptors.

the boundary conditions for this problem are presented in Table 1. For humans, major potential routes include: inhalation of dusts, aerosols and vapors; ingestion of contaminated fish, foodstuffs and soil; ingestion of contaminated

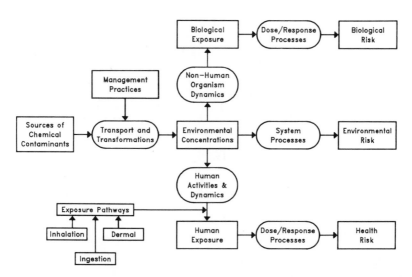

Figure 2.   Concept of exposure and risk assessment.

drinking water; and dermal absorption from air, water or soil.   For fish, the exposure routes include contact with ambient waters and sediments and bioaccumulation. Similarly, wildfowl exposure routes include ambient contact, bioaccumulation and direct ingestion.   The interrelationships between sources, pesticide BMPs and exposure concentration to biological, environmental and human health risks is seen in Figure 2.

Any serious attempt to assess the exposure to pesticides used in conservation tillage should include all application-related sources of pesticides and routes of exposure (Figure 3) and properly treated on-site exposures (e.g., birds in agricultural fields) and off-site exposures (e.g., fish in drainage waters).   To simplify the problem, however, we have chosen to concentrate on off-site concerns (e.g., drinking water contamination via leaching) thereby reducing the analysis problem to a careful look at the pesticide fluxes derived from conservation tillage systems.   Thus, we can immediately examine questions concerning the impact of conservation tillage on pesticide runoff (dissolved and particulate-bound forms), leaching and volatilization pathways.

The basis for any quantitative analysis of fluxes must be the mass balance approach.   A qualitative description of the conservation tillage system can be written as:

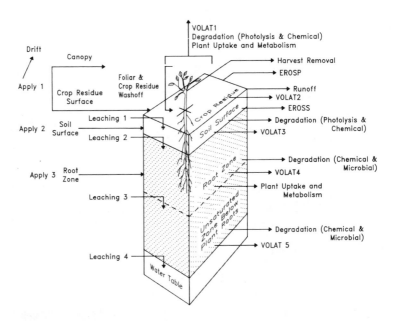

Figure 3. Pesticide-soil-plant mass balance.

"SOURCE"   "SINKS"

PESTICIDE APPLICATION = (DEGRADATION LOSSES + VOLATILIZATION
                        LOSSES + EROSION RUNOFF + LEACHING
                        + HARVEST REMOVAL + Δ STORAGE)

Our basic problem is to understand how each of the above terms is influenced by conservation tillage. It is convenient to further describe the conservation tillage system as having four interacting components with a mass balance written for each.

Consider such a system as depicted in Figure 3.

For the plant canopy and crop residue we can write:

APPLY1 = DEGRADATION (PHOTO + CHEMICAL) + VOLAT1
         + PLANT UPTAKE + HARVEST + PLANT METABOLISM
         + FOLIAR AND RESIDUE WASHOFF + EROSP
         + VOLAT2 - VOLAT3 + Δ STORAGE

Next, for the soil surface zone we can write:

APPLY2 = DEGRADATION (PHOTO + CHEMICAL) + VOLAT3
        + EROSS + RUNOFF + LEACHING1 - VOLAT4
        - FOLIAR/RESIDUE WASHOFF + Δ STORAGE

For the root zone we have:

APPLY3 = DEGRADATION (CHEMICAL + BIOLOGICAL)
        + VOLAT4 - VOLAT5 + LEACHING3 - LEACHING2
        + PLANT UPTAKE + Δ STORAGE

Finally, for the zone below the root zone we have:

Δ STORAGE = LEACHING3 - LEACHING4
        - DEGRADATION (CHEMICAL) - VOLAT5

Note in the above equations that some processes are internal
to the system. We are more interested in the environmental
fluxes out of the system for our exposure analysis--specifically,
VOLAT1, HARVEST, EROSP, VOLAT2, EROSS, RUNOFF, VOLAT3
and LEACHING2 and 4. Recall that VOLAT1, VOLAT2 and VOLAT3
represent vapor losses from the plant canopy, surface
residue and soil surface, respectively. EROSP, EROSS
and RUNOFF represent erosion losses in plant residue and
soil and runoff losses in overland flow, respectively.
LEACHING2 represents leaching below the root zone and
HARVEST represents removal via plant burdens.

Consider a mass-balance equation for the soil profile
from the surface to the bottom of the root zone derived
in terms of the soluble pesticide form, C_w. It is readily
shown that a reasonable approximation of this system is:

$$\frac{\partial C_w}{\partial t} = \frac{(D_a + D_w)(\theta_{fc} + K_H(1-\theta-\rho_s-\rho_p))}{\epsilon} \cdot \frac{\partial^2 C_w}{\partial z^2} - \frac{V_w\theta + V_a K_H(1-\theta-\rho_s-\rho_p)}{\epsilon} \cdot \frac{\partial C_w}{\partial z} -$$

$$\frac{\left[\Sigma k_w\theta_{fc} + \Sigma k_s K_D^s \rho_s + \Sigma k_p K_D^p \rho_p + \Sigma k_a K_H(1-\theta_{fc}-\rho_s-\rho_p)\right] C_w}{\epsilon} -$$

$$\frac{(EROSS \cdot K_D^s \rho_s + EROSP \cdot K_D^p \rho_p + kT + RUNOFF) C_w}{\epsilon}$$

where: C_w = soluble pesticide concentration
        D_a, D_w = diffusion-dispersion coefficients
                for air and water
        θ = volumetric water content
        θ_{fc} = volumetric water content at field
                capacity
        K_H = Henry's constant

$$\rho_s, \ \rho_p = \text{soil and plant residue density}$$

$$K_D{}^S, \ K_D{}^P = \text{partition coefficients for soil and plant residue}$$

$$V_W, \ V_a = \text{velocity vectors for water and air}$$

$$\Sigma k_w, \Sigma k_s, \Sigma k_p, \Sigma k_a = \text{sum of first-order rate constants for degradation in water, on soil, on plant residue, and in air}$$

$$EROSS = \text{soil erosion rate}$$

$$EROSP = \text{plant residue erosion rate}$$

$$kT = \text{transpiration rate of contaminated soil water}$$

$$RUNOFF = \text{water runoff rate}$$

$$z = \text{vertical depth dimension}$$

and finally,

$$\epsilon = \theta_{fc} + K_D^p \rho_p + K_D^s \rho_s + K_H (1 - \theta_{fc} - \rho_s - \rho_p)$$

The above equation is useful because it provides a means to isolate major system variables influenced by conservation tillage and examine the nature of that impact. Note that the environmentally relevant variables are θ_{fc}, V_w, V_a, ρ_s, ρ_p, EROSS, EROSP, and RUNOFF. These variables are not independent in every case, and in some cases their dependence may be highly non-linear. For example, $V_w = f(\theta_{fc}, RUNOFF, \rho_s)$. The chemical-specific variables include D_a, D_w, K_H, K_D^P, K_D^S, Σk_s, Σk_p, Σk_w, and Σk_a. Here again, these values may be dependent and related in a non-linear fashion. The point is that the above equation enables a ranking of impacts if we know the impact of conservation tillage on the relevant variables.

Trends in chemical properties give a clear indication of the impact of chemical properties on the resulting fluxes. For example, as K_D^P and K_D^S increase, then EROSS, EROSP and RUNOFF decrease. If they decrease, then the dispersive and advective terms (the first two in the equation) increase with the attendant increase in leaching. Additional observations can be noted once a clearer picture of the relative impact of conservation tillage on each variable becomes apparent.

It is useful to rank the impact that conservation tillage has on each of the terms in the above mass balance equations. Table 2 identifies the major fluxes of concern and the off-site exposure routes and presents a judgmental ranking of the impact of conservation tillage relative to conventional systems. Some interesting (but not surprising)

Table 2. Influence of conservation tillage on mass balance variables.

Zone of Interest Plant Canopy and Surface Residue	Relative Change from Conventional Tillage	Flux of Concern to Off-Site Exposure	Pollution Concern
APPLY 1			
DEGRADATION	large increase		
VOLAT1	unknown	x	air
PLANT UPTAKE	large increase	x	foodstuff
HARVEST	no impact		
FOLIAR/RESIDUE WASHOFF	small increase	x	surface water
EROSP	large increase		surface water
VOLAT2	large increase	x	air
VOLAT3	large increase	x	air
SOIL SURFACE			
APPLY 2			
DEGRADATION	small increase		
VOLAT3	small reduction	x	air
EROSS	large reduction	x	surface water
RUNOFF	large reduction	x	surface water
LEACHING1	large increase		groundwater
VOLAT4	large reduction		air
FOLIAR/RESIDUE WASHOFF	large increase		
ROOT ZONE			
APPLY 3			
DEGRADATION	no impact		
VOLAT4	large reduction		air
LEACHING2	large increase	x	groundwater
PLANT UPTAKE	no impact		foodstuff

observations are readily obvious. Atmospheric losses of pesticides are likely to increase because more are applied to plant, soil and residue surfaces. Runoff and soil-erosion-related losses will decrease but plant-residue-associated "erosion" will increase. Leaching will increase.

The above qualitative ranking has served to narrow the problem somewhat and indeed the major concern appears to be the tradeoff between runoff and the resulting impacts on various receptors in surface water and the leaching losses to groundwater. Another potential concern is the loss of pesticides from crop residue through the volatilization pathway. Research is needed to evaluate this tradeoff in more detail.

ASSESSMENT OF THE IMPACT OF CONSERVATION TILLAGE
ON SOIL PROPERTIES, PROCESSES AND PESTICIDE
TRANSPORT AND TRANSFORMATION

In the previous section we developed an approach to identify major system variables influenced by conservation tillage and speculated on the impact on each by conservation tillage. In this section, we will attempt to document from the literature our speculations on the impact of conservation tillage on these relevant variables (Table 3).

Transformation Processes

Few actual data exist on pesticide transformation rates in soils subjected to conservation vis-a-vis conventional tillage. Some implications of likely behavior can be inferred, however, from the documented changes in soil properties. Specifically, the increase in organic matter levels (Dick, 1983; Blevins et al., 1983; Langdale and Leonard, 1983; Langdale et al., 1984) and the increase in microbial activity of the upper soil layers (Doran, 1980) combined with a higher average moisture content, less rapidly changing and slightly lower temperatures (Gupta et al., 1982), and the possibility of improved oxygen transport in the upper soil layers due to reduced bulk density and increased bioturbation should increase the overall effective rates of both biological and chemical oxidation. Indeed, both aerobic (initially), facultative and anaerobic activity may later be increased as a result of the higher moisture levels and initially rapid biological degradation of the organic residue. In addition, the elevated moisture levels should increase mass transfer and diffusion rates of the pesticides to, into and from the soil such that soil-mediated reactions are also accelerated.

The presence of large quantities of organic matter on or in the upper soil profile combined with increased biological activity should produce acidic conditions that, in turn, will possibly require increased lime and ammonia additions for both pH and nutrient adjustments. As a result, it is likely that pesticides in the upper soil zones will be subjected to wider swings in pH than present under conventional tillage, and therefore, both acid and

Table 3. Impact of conservation tillage on soil properties, activities, and processes.

- - - -Property/Activity- - - -		- - - -Processes- - - -	
Increase	Decrease	Increase	Decrease
Organic matter	soil temp.	infiltration	runoff
Microbial population	pH	hydraulic conductivity	water erosion
Soil aggregation	exchangeable calcium and magnesium	nitrogen immobili- zation	evaporation
Bioturbation			wind erosion
Soil structure	nitrate- nitrogen	leaching	
Exchangeable aluminum and manganese	bulk density*		
Water storage			
Bulk density*			
Porosity			
Available phosphorus and potassium			
Organic nitrogen			

*Function of soil type and climatic conditions.

base hydrolysis reactions should be accelerated. Research results confirm that soil pH does decrease with conservation tillage (Dick, 1983; Blevins et al., 1983). With regard to transformation in the lower soil zones, there is not enough information to draw any definitive conclusions, but it would appear that there should be no significant difference as compared to conventional tillage. A paucity of information exists on pesticide transformation rates in the lower horizons.

The real unknown in this analysis is the net effect of the vastly increased surface area of organic matter left on the soil surface relative to pesticide transformation pathways and rates. Even though most "new" pesticides have low partition coefficients, the large amount of organic matter present combined with its high surface area and the generally higher pesticide usage for conservation tillage would imply a potentially pivotal role for this material in pesticide degradation and transport. Nominally, one would assume that pesticides intercepted or sorbed on this organic surface could undergo photolysis reactions to a much greater degree and for a longer time than soil-applied chemicals in conventional tillage. Actual quantification of this hypothesized role is an urgent research need.

Pesticide Transport to Other Environmental Media

In this section, we will briefly examine the effect of conservation tillage on the four major transport pathways—runoff, erosion, leaching and volatilization—on the intermedia transfer of pesticides to surface water, to groundwater and to the atmosphere. A summary is given in Table 1.

Runoff. Conservation tillage reduces the total amount of runoff reaching streams, i.e., reduces runoff volume (Harold and Edwards, 1974; Langdale et al., 1978; Langdale et al., 1979; Thomas et al., 1982; Burwell and Kramer, 1983; Langdale et al., 1983). The extent is a function of such factors as the specific type of conservation system, soil properties, slope, amount of crop residue, crop type, percentage of surface coverage and antecedent soil moisture among other factors. For example, Langdale and coworkers (1979) found that, in a comparison of conventional versus no-till, double-cropped barley followed by grain sorghum with fescue, annual runoff was decreased from 22.8 cm to 12.0 cm. More potentially important from an environmental standpoint was that runoff was reduced 90 percent from May through July, during which time both sediment, sediment-bound and dissolved pesticide transport potential is usually the highest. Conservation tillage protects the surface of the soil at a time when the soil is the most susceptible to erosion and pesticide runoff.

Pesticides of varying physical, chemical and toxicological characteristics have been found in runoff; the data has been summarized by Leonard and coworkers (1976), Wauchope (1978) and by Baker and Johnson (1983). Factors that influence runoff (and erosion) and thus pesticide transport in overland flow include land management and conservation practices, rainfall characteristics, land usage and soil erodibility.

Little field research has been reported comparing pesticide losses in runoff between conservation and conventional tillage. Consequently, few data exist to parameterize either water balance or pesticide transformation properties under conservation tillage for modeling purposes. Baker and Johnson (1979) found that fonofos was lost in greater amounts from till-plant systems than from either conventional or ridge plant systems; the greater part of losses being in runoff.

In general, for those pesticides with low or moderate K_p or K_{oc}, the major mode of transport is in runoff rather than delivery in sediment. The converse is true for pesticides with high K_p or K_{oc}. The concentration is higher on the sediment, but the greater load is carried by runoff because of the greater quantity of water than soil transported by overland flow. For example, in a study on these same watersheds (Smith et al., 1978), it was found that six herbicides (atrazine, cyanazine, diphenamid, trifluralin, propazine, and 2,4-D) showed higher concentrations on sediment than in runoff, but the total mass of pesticide transported in the water phase was much greater than in the sediment phase because of the greater mass of water compared to the mass of sediment.

Caution must be exercised when reviewing the literature because broad, sweeping generalizations have been made that the major transport route is pesticide bound on sediment rather than dissolved in runoff. These conclusions were based on early results for the persistent, chlorinated hydrocarbons with high K_p's. Insecticides and herbicides today generally have much lower K_p and K_{oc} (< 1000 compared to the chlorinated hydrocarbons, $K_{oc} > 10,000$). They also are much less persistent in soils.

Because, in general, the application rates of pesticides for conservation tillage are higher than for conventional tillage systems, the potential concentration to be lost via overland flow is greater. This fact, all things equal, would be significantly more important for those rainfall-runoff producing events occurring on or shortly after the day of application.

More sources for contribution to runoff as defined in Figure 3 are available under conservation compared to conventional tillage. Foliar washoff and crop residue washoff, as well as "extraction" from the soil surface, contribute to runoff and overland flow to surface water. The contribution of each source to the total pesticide load being delivered in runoff via overland flow is not known, and no field studies to quantify these loads are underway. The affinity or retention characteristics of foliar and crop residue surfaces remains to be determined. How different it is from the partitioning estimates based on organic carbon content of soils also remains to be determined.

Conservation tillage decreases runoff and increases infiltration by both decreasing the formation of surface crusting and by increasing and maintaining macropores in the surface horizon. This is due to increased organic matter content and bioturbation in the plant root zone.

Erosion. Erosion may be caused by both water and wind, and the erosional processes may generate both soil particles and organic detritus. Both of these substrates may be vehicles to transport pesticides (and nutrients as well) off-site to receiving surface water or some other receptor. Wind-blown dusts have been found to be carriers for chlorinated hydrocarbon insecticides (Cohen and Pinkerton, 1966), and such fugitive dusts have been transported hundreds or thousands of miles from their sources.

Conservation tillage reduces water erosion and soil losses compared to conventional tillage (Mannering and Burwell, 1968; Harold and Edwards, 1974; Langdale et al., 1978; Fenster and McCalla, 1970; Fenster, 1977; Reicosky et al., 1977). For example, Langdale and coworkers (1978) found a reduction in soil erosion in the Southern Piedmont from 26.26 metric tons/ha (11.71 tons/acre) with conventional tillage to 0.131 metric tons/ha (0.06 tons/acre) with no-tillage. No chemical data were given so direct comparison could not be made on the impact of tillage systems on the flux of pesticides loss from the small upland watersheds.

Although soil loss is reduced many fold by the use of conservation tillage, caution must be exercised in ruling out completely the route of water erosion as an important transport route. Erosion is a selective process whereby the finer particles—clay and organic matter (organic matter moved as a particulate or bound onto inorganic substrates)—are delivered to streams. Because sorption capacity is highest with the smaller particle size and because organic matter has been found to be best correlated with sorption for most classes of pesticides, these two fractions may be "hotter" than the whole soil.

Water erosion may detach and transport crop residue detritus from the field and deliver it to surface water. Pesticides bound on crop residue may, therefore, enter surface water in this manner. The importance of this transport route, the ease of desorption from the crop residue and the flux/loads generated from this source are not known presently.

Leaching and Groundwater. Conservation tillage, as stated earlier, reduces the volume of runoff by an average of 25 percent (Baker and Johnson, 1983); the exact degree of reduction is highly variable and site-specific. Conservation tillage favorably influences those physical soil properties that increase infiltration and overall water storage. Under conservation tillage, organic matter content, soil aggregation and soil structure are increased; the continuity of macro-channels in soil is maintained, thus permitting better communication to subsoil; bioturbation near the surface is increased; and overall bioactivity is increased compared to conventional tillage systems (Edwards, 1982; Unger and McCalla, 1980; Klute, 1982; Larson and Osborne, 1982; Smith and Rice, 1983). The property, bulk density, is increased or decreased depending upon soil type and climatic conditions. The environmental properties that most influence the potential of a pesticide to leach to subsurface waters (interflow) or to groundwater are its persistence and mobility through soil and subsurface porous media. Persistence is a collective term representing a chemical's overall behavior with respect to degradation processes and volatilization. Mobility is related to the pesticide distribution coefficient and water solubility of the pesticide.

Certain pesticides have been found in tile drainage water. Gold and Loudon (1982) reported atrazine at a concentration range of 20 to 170 ppb in tile drainage water generated by a storm four days after planting. Of the 1.4 kg/ha applied at planting, 1.0 and 2.2 percent were lost in tile flow from conventional and conservation tillage, respectively. Overland flow concentrations were two to four times higher than those found in tile flow.

Pesticides have been found in groundwater (Harkin et al., 1982; Hebb and Wheeler, 1978; Peoples et al., 1980; Rothschild et al., 1982; Spalding et al., 1979; Wehtze et al., 1983; Zake et al., 1982; Cohen et al, 1984). Cohen and coworkers (1984) reported that a total of 12 pesticides had been found in the groundwater of 18 different states. These include alachlor, aldicarb (sulfoxide and sulfone), atrazine, simazine, bromacil, carbofuran, dacthal, DBCP, 1,2,dichloropropane, dinoseb, EPB and oxamyl. The majority of these fall into three categories--triazines, carbamates and low molecular weight hydrocarbons. In many cases, these herbicides/nematicides were found in groundwater below sandy soils under intense irrigation.

Groundwater concentration levels were in the medium parts-per-trillion to the low parts-per-billion range.

Volatilization. Pesticides applied to crops may reside in or on one or more surfaces (Figure 3)--crop canopy, crop residue and soil surface. The pesticides may ultimately be incorporated into the soil following the initial rainfall or by mechanical incorporation. Tillage practices may influence the distribution between the crop residue surface and the soil surface. In the case of conservation tillage, the surface crop residue acts as a physical barrier preventing pesticides from reaching the soil surface, particularly broadcast sprayed herbicides. Several review articles summarize research on chemical, environmental and soil factors influencing vapor transport and loss of pesticides from both soil incorporated and surface applied pesticides (Spencer, 1970; Spencer et al., 1973; Spencer and Claith, 1977; Spencer et al., 1982; Guenzi and Beard, 1974; Farmer et al., 1984). Little is known from a predictive sense about the magnitude of vapor losses from foliage-applied pesticides. Worker reentry times for foliar-applied insecticides is a public health concern.

The major difference between conservation tillage and conventional tillage from a pesticide transport and export standpoint is that pesticides are applied mainly to the crop residue surface in the former case and directly on or incorporated into the soil in the latter case. This is a major difference because herbicides, in particular, are normally applied as a spray and not as a granule.

Concern has been expressed (Martin et al., 1978; Baker, 1980; Baker and Laflen, 1983; Baker, 1985) about herbicide volatilization from surface crop residue, but the magnitude of the problem has not been quantified. We do not know whether the herbicide is adsorbed or absorbed by the crop residue. We also do not know whether the chemical can be desorbed from crop residue surfaces or diffused to the crop residue surface from its interior and lost to the atmosphere. One would expect higher volatility losses from chemicals that have a high vapor pressure and a low partition coefficient. Picloram was not found to be adsorbed by wheat straw (Grover, 1971).

With surface temperatures rising to 50 to 60°C at times, volatilization from crop residue may be a major transport pathway from the crop/soil system to the atmosphere. Both on- and off-site human impacts will be determined by the initial concentration of the pesticide, flux from the crop residue, photostability, drift characteristics and inhalation toxicity of the pesticide.

If pesticide volatilization from the crop residue is a major environmental transport pathway, then improvements in application technology may be needed, e.g., use of granular formulation and/or application technology whereby the residue is disturbed and the pesticide (whether spray or granular form) is applied mainly to the soil surface. The presence of crop residue then would decrease pesticide loss by increasing the diffusion path length to the atmosphere, by lowering the soil surface temperature and by decreasing the wind velocity at the soil surface.

ENVIRONMENTAL IMPLICATIONS

It is not clear either from model simulations or actual data what the net environmental impact will be from pesticides that are applied in conjunction with the widespread adoption of conservation tillage. Although it is clear that both surface runoff water and sediment volumes will be reduced and that many degradation processes might be accelerated, it is possible that the increased amounts and frequency of pesticide use will offset this apparent benefit.

Clearly, the potential for direct exposure to farm workers and rural residents will be increased due to higher pesticide rate of nongranular formulations, enhanced volatilization losses and off-site drift from aerial applications. Similarly, the decreased surface runoff will result in a general increase in potential threats to groundwater from mobile, soluble and persistent pesticides. Fines enrichment and the increase in organic particulate runoff could result in sufficient sorbed pesticide at higher concentrations such that the total sediment-bound loads could exceed those of conventional tillage on the same fields. Similarly, even though the runoff water volumes are reduced, higher concentrations of pesticides in it could produce more acute impacts in receiving waters and even greater total loads than that resulting from conventional operations.

One potentially serious environmental impact that needs to be evaluated, although not related to pesticides, is a direct result of the projected major decrease in sediment loadings to surface streams. Under such drastic changes in load, data (Simons and Li, 1980) suggest that stream bed and bank erosion rates may be greatly increased, thereby seriously threatening the habitats of upland stream and riverine aquatic organisms. Moreover, a major reduction in average instream sediment concentrations without a concurrent major reduction in nutrient and carbon loads to surface waters could aggravate the tendency toward stream/river algae blooms produced by an increase in light penetration ability.

Obviously, targeted field studies are needed to quantify these various environmental concentrations and fluxes, to verify the projected changes and to parameterize assessment models for subsequent regional and national analyses of water quality and human exposure impacts likely to result form the widespread adoption of conservation tillage.

Conservation tillage increases infiltration and hydraulic conductivity, which would favor the leaching of pesticides. Increased organic levels in surface soils would promote the retention of those pesticides having a moderate K_p, K_{oc}. The root zone environment would experience higher biological activity, and a different distribution of microorganisms would be found (Doran, 1980). Aerobic microorganism counts increased 10 to 80 percent in the 0 to 7.5 cm depth, and anaerobic bacteria increased 60 to 300 percent in the reduced tillage plots compared to conventional tillage. In the 7.5 to 15 cm depth, populations of facultative anaerobes and denitrifiers were higher with no-till. This indicates that conservation tillage soils are less oxidative than conventional tillage soils.

The significance of these different types of microbial populations as a function of tillage systems on the relative resistance of a pesticide to degradative processes is unclear at this time. The nature of degradative environments in terms of both biological and chemical processes as a function of the depth in the profile and in the porous media above the groundwater aquifer needs to be defined. Appropriate kinetic expressions must be developed to quantitatively describe degradation processes all the way to the aquifer.

SUMMARY AND CONCLUSIONS

Conservation tillage is projected to be the major soil protection method and candidate best management practice for improving surface water quality. Conversion from conventional to conservation tillage practices is proceeding at an accelerating rate. Specifically, it is estimated that by the turn of the century about 95 percent, or 162 million of the 170 million hectares devoted to row crops in the United States will be cultivated by conservation tillage. This quantum level change in management systems is comparable to that which occurred in the 1950s when rotation-based agriculture was changed to synthetic-chemical-based agriculture.

Benefits of conservation tillage as brought out by papers at this conference include: reduced erosion, lower sediment-associated pesticide losses to surface water, increased infiltration of water into the soil, reduced runoff, higher moisture content in the surface horizon, reduced energy use and costs and improved long-term soil and crop productivity.

Conservation tillage also raises environmental concerns because of increased pesticide application rates, lack of soil incorporation of pesticide and potentially enhanced pesticide volatilization fluxes to the atmosphere.

Potential environmental implications of conservation tillage characteristics that are mainly off-site in nature include: (1) increased direct exposure to farm workers and rural residents due to enhanced volatilization, i.e., greater acute and chronic impact; (2) increased threat of groundwater pollution (greater infiltration and leaching of soluble and persistent pesticides), i.e., enhanced adverse impacts on aquifers for drinking water and irrigation water supplies; (3) greater total sediment-bound pesticide loads (due to clay particle size enrichment) and greater instream impact on sensitive ecological systems; (4) greater pesticide concentrations in runoff (nonsoil incorporation of pesticides) and greater instream impact on sensitive ecological systems; (5) decreased stream bed and bank stability (decreased total erosion and soil loss and increased stream scour carrying capacity) with riparian habitats threatened; and (6) reduced stream and lake sediment concentrations resulting in increased light penetration and enhanced algal bloom.

One approach to fully sustain the extensive implementation and use of conservation tillage and obviate the potential adverse environmental implications would be the periodic use of primary or conventional tillage practices and adequate crop rotation. For example, in the Southeast the use of primary tillage once every three to five years and rotation of the cash crop, i.e., soybeans and corn, appears to be one feasible alternative[1]. Primary tillage would be done in the fall preceding the sowing of the double crop, wheat; the crop would provide cover to reduce erosion and runoff during the period of intense rainfall that occurs before conservation tillage practices are begun.

---

[1]Personal communication, George W. Langdale.

Answers to questions about the environmental impact and implications of nationwide acceptance and implementation of conservation tillage as a best management practice will require incisive, innovative field and laboratory research to quantify water balance components and transformation rates and to parameterize assessment models subsequent to regional, national and site-specific assessments.

LITERATURE CITED

Bailey, G.W. and R.R. Swank, Jr. 1983. Modeling agricultural nonpoint source pollution: A research perspective. In: Agricultural Management and Water Quality, F.W. Schaller and G.W. Bailey (eds.), Iowa State Univ. Press, Ames, IA.

Bailey, G.W., R.R. Swank, Jr., and H.P. Nicholson. 1974. Predicting pesticide runoff from agricultural land: A conceptual model. J. Environ. Qual. 3:95-102.

Bailey, G.W. and T.E. Waddell. 1979. Best management practices for agriculture and silviculture: An integrated overview. In: Best Management Practices for Agriculture and Silviculture, R.C. Loehr, D.A. Haith, M.F. Walter, and C.S. Martin (eds.), Ann Arbor Sci. Pub., Inc., Ann Arbor, MI.

Baker, J.L. 1980. Agricultural areas as nonpoint sources of pollution. In: Environmental Impact of Nonpoint Sources of Pollution, M.R. Overcash and J.M. Davidson (eds.), Ann Arbor Sci. Pub., Inc., Ann Arbor, MI.

Baker, J.L. 1985. Conservation tillage: Water quality considerations. In: A Systems Approach to Conservation Tillage, F.M. D'Itri (ed.), Lewis Publishers, Inc., Chelsea, MI 48118, pp 217-238.

Baker, J.L. and H.P. Johnson. 1979. The effect of tillage systems in pesticide runoff from small watersheds. Trans. Am. Soc. of Agric. Engrs., pp 554-559.

Baker, J.L. and H.P. Johnson. 1983. Evaluating the effectiveness of BMPs via field studies. In: Agricultural Management and Water Quality, F.W. Schaller and G.W. Bailey (eds.), Iowa State Univ. Press, Ames, IA.

Baker, J.L. and J.M. Laflen. 1983. Water quality consequences of conservation tillage. J. Soil Water Cons. 33:186-191.

Blevins, R.L., N.S. Smith, G.W. Thomas, and W.W. Fyre. 1983. Influence of conservation tillage on soil properties. J. Soil Water Cons. 38:301-305.

Burwell, R.E. and L.A. Kramer. 1983. Long term annual runoff and soil loss from conventional and conservation tillage of corn. J. Soil Water Cons. 38:315-319.

Cohen, J.M. and C. Pinkerton. 1966. Widespread translocation of pesticides by air transport and rain-out. In: Organic Pesticides in the Environment, Advances in Chemistry Series 60, American Chemical Society, Washington, DC.

Cohen, S.Z., S.M. Creeger, R.F. Carsel, and C.G. Enfield. 1984. Potential pesticide contamination of groundwater from agricultural uses. In: Treatment and Disposal of Pesticide Wastes, R.F. Kruger and J.N. Sieber (eds.), ACS Sym. Series No. 259, American Chemical Society, Washington, DC.

Crosson, P. 1981. Conservation tillage and conventional tillage: A comparative assessment. Soil Conserv. Soc. Amer. Monogr., Ankeny, IA.

Crosson, P. and S. Brubaker. 1982. Resource and Environmental Effects of U.S. Agriculture. Resources for the Future, Inc., John Hopkins Univ. Press, Baltimore, MD.

Crosson, P. 1983. Trends in agriculture and possible environmental futures. In: Agriculture Management and Water Quality, F.W. Schaller and G.W. Bailey (eds.), Iowa State Univ. Press, Ames, IA.

Dick, W.A. 1983. Organic carbon, nitrogen, and phosphorus. Soil Sci. Soc. Amer. J. 47:102-107.

Doran, J.W. 1980. Soil microbial and biochemical changes associated with reduced tillage. Soil Sci. Soc. Amer. J. 44:765-771.

Edwards, P.M. 1982. Predicting tillage effects on infiltration. In: Predicting Tillage Effects in Soil Physical Properties and Processes. ASA Special Publication Number 44:105-115, P.W. Unger and D.M. Van Doren, Jr. (eds.), Madison, WI.

Farmer, W.J., W.F. Spencer, and W.A. Jury. 1984. Soil processes and their use in predicting volatilization of pesticides from soil. In: Prediction of Pesticide Behavior in the Environment: Proceedings of US-USSR Symposium, October, 1981, Yerevan, USSR, p. 110-123, U.S. Environmental Protection Agency, Athens, GA. EPA 600/9-84-026.

Federal Water Pollution Control Acts of 1972. Public Law 92-500. October, 1972.

Federal Water Pollution Control Act of 1977. Public Law 95-217. December, 1979.

Fenster, C.R. 1977. Conservation tillage in the Northern Plains. J. Soil Water Cons. 32:37-42.

Fenster, C.R. and T.M. McCalla. 1970. Tillage practices in Western Nebraska with a wheat-fallow rotation. Publication 597, Nebraska Agriculture Experiment Station, Lincoln, NE.

Gold, A.J. and T.L. Loudon. 1982. Nutrient, sediment, and herbicide losses in tile drainage under conservational and conventional tillage. Am. Soc. of Agric. Engineers. National Meeting. Chicago, IL, Paper No. 82-2549.

Grover, R. 1971. Adsorption of picloram by soil colloids and various other adsorbents. Weed Sci. 19:417-418.

Guenzi, W.D. and W.E. Beard. 1974. Volatilization of pesticides. In: Pesticides in Soil and Water. W.D. Guenzi (ed.), Soil Science Society of America, Inc., Madison, WI.

Gupta, S.C., W.E. Larson, and D.R. Linden. 1982. Tillage and surface residue effects on soil upper boundary temperature (Zea mayes). Soil Sci. Soc. Amer. J. 74:1212-1218.

Harkin, J.M., F.A. Jones, R. Fathulla, E.K. Dzantor, and E.J. O'Neil. 1984. Pesticides in groundwater beneath the Central Sand Plain of Wisconsin. Wisconsin Univ., Madison Water Resources Center, Madison, WI.

Harold, L.L. and W.M. Edwards. 1974. No-tillage systems reduce erosion from continuous corn watersheds. Trans. Amer. Soc. Agric. Engr. 17:414-416.

Hebb, E.A. and W.B. Wheeler, 1978. Bromacil in Lakeland soil and groundwater. J. Environ. Qual. 7:498-601.

House, G.J., B.R. Stinner, D.A. Crossley, Jr., E.P. Odum, and G.W. Langdale. 1984. Nitrogen cyclng in conventional and no-tillage agroecosystems in the Southern Piedmont. J. Soil Water Cons. 39:194-200.

Klute, A. 1982. Tillage effects on the hydraulic properties of soil: A review. In: Predicting Tillage Effects on Soil Physical Properties and Processes, P.W. Unger and

D.M. Van Doren, Jr. (eds.), Amer. Soc. Agron. Monogr., No. 44.

Langdale, G.W., A.P. Barnett, and J.E. Box, Jr. 1978. Conservation tillage systems and their control of water erosion in the Southern Piedmont. Proceedings of the First Annual Southeastern No-Till Systems Conference. J.T. Touchton and D.G. Cummings (eds.), Special Publication No. 5:22-29.

Langdale, G.W., A.P. Barnett, R.A. Leonard, and W.G. Fleming. 1979. Reduction of soil erosion by the no-till system in the Southern Piedmont. Trans. Am. Soc. Agric. Engr. 22:82-92.

Langdale, G.W., W.L. Hargrove, and J. Giddens. 1984. Residue management in double-crop conservation tillage systems. Agron. J. 76:689-694.

Langdale, G.W. and R.A. Leonard. 1983. Nutrient and sediment losses associated with conventional and reduced tillage agricultural policies. In: Nutrient Cycling in Agricultural Ecosystems, R.L. Todd, L. Amussen, and R.A. Leonard (eds.), Univ. of Georgia, College of Agriculture Experiment Station, Special Publication 23, pp. 457-467.

Langdale, G.W., H.F. Perkins, A.P. Barnett, J.C. Readon, and R.L. Wilson, Jr. 1983. Soil and nutrient runoff losses within-row chisel-planted soybeans. J. Soil Water Cons. 38:297-301.

Larson, W.E. and G.J. Osborne. 1982. Tillage accomplishments and potential. In: Predicting Tillage Effects of Soil Physical Properties and Processes, P.W. Unger and D.M. Van Doren, Jr. (eds.), Amer. Soc. Agron. Monog. No. 44.

Leonard, R.A., G.W. Bailey, and R.R. Swank, Jr. 1976. Transport, detoxification, fate and effects of pesticides in soil and water environments. In: Land Application of Waste. Soil Conserv. Soc. Amer., Ankeny, IA.

Mannering, J.V. and R.E. Burwell. 1968. Tillage methods to reduce runoff and erosion in the corn belt. Bull. 330, USDA, Washington, DC.

Martin, L.D., J.L. Baker, D.C. Erback, and H.P. Johnson. 1978. Washoff of herbicides applied to corn residues. Trans. Amer. Soc. Agric. Engr. 21:1164-1168.

Peoples, S.A., K.T. Maddy, W. Cusick, T. Jackson, C. Cooper, and A.S. Frederickson. 1980. A study of samples of well water collected from selected areas in California to determine the presence of DBCP and certain other pesticide residues. Bull. Environ. Contamin. Toxicol. 24:611-618.

Reicosky, D.C., D.K. Cassell, R.L. Blevein, W.R. Gillard, G.C. Naderman. 1977. Conservation tillage in the Southeast. J. Soil Water Cons. 32:13-19.

Rothschild, E.R., R.L. Manser, and M.J. Anderson. 1982. Investigation af aldicarb in groundwater in selected areas of the Central Sand Plain of Wisconsin. Groundwater. 20:437-445.

Simons, D.B. and R.M. Li. 1980. Modeling of sediment nonpoint source pollution from watersheds. In: Environmental Impacts of Nonpoint Source Pollution. M.R. Overcash and J.M. Davidson (eds.), Ann Arbor Sci. Publ., Ann Arbor, MI.

Smith, C.N., R.A. Leonard, G.W. Langdale, and G.W. Bailey. 1978. Transport of Agricultural Chemicals from Upland Piedmont Watersheds. U.S. Environmental Protection Agency, Athens, GA. EPA-600/3-78-056.

Smith, M.S. and C.W. Rice. 1983. Soil biology and biochemical nitrogen transformation in no-tilled soils. In: Environmentally Sound Agriculture. W. Lockeretz (ed.), Praegh Scientific.

Spalding, R.F., G.A. Junk, and J.J. Richards. 1980. Water-pesticides in groundwater beneath irrigated farmland in Nebraska. Pest. Monit. J. 14:70-73.

Spencer, W.F. 1970. Distribution of pesticides between soil, water, and air. In: Pesticides in the Soil: Ecology, Degradation and Movement. Michigan State Univ., East Lansing, MI, pp. 120-128.

Spencer, W.F. and M.M. Claith. 1972. The solid-air interface: Transfer of organic pollutants between the solid-air interface. In: Fate of Pollutants in the Air and Water Environments: Part I. I.H. Suffett (ed.), Wiley, New York, NY.

Spencer, W.F., W.J. Farmer, and M.M. Claith. 1973. Pesticide volatilization. Residue Reviews. 49:1-47.

Spencer, W.F., W.J. Farmer, and W.A. Jury. 1982. Review: Behavior of organic chemicals at soil, air, water interfaces as related to predicting the transport and volatilization of organic pollutants. Environ. Tox. Chem. 1:17-26.

Thomas, S.W., G.W. Langdale, and E.L. Robinson. 1982. Tillage and double-cropped practices on watershed. In: Proceedings of Speciality Conference on Environmentally Sound Water and Soil Management. E.G. Kruse and Y.A. Yousef (eds.), Am. Soc. Civil Engr., New York, NY.

Unger, P.W. and T.M. McCall. 1980. Conservation tillage systems. Advances in Agronomy. Nyle C. Brady (ed.), Amer. Soc. Agron., Madison, WI.

Unger, S.G. 1977. Environmental Implications of Trends in Agriculture and Silviculture, Vol. 1. Trend Identification and Evaluation. U.S. Environmental Protection Agency, Athens, GA. EPA-600/3-77-121.

Unger, S.G. 1978. Environmental Implications of Trends in Agriculture and Silviculture, Vol. 2. Environmental Effects of Trends. U.S. Environmental Protection Agency, Athens, GA. EPA-600/3-78-102.

Unger, S.G. 1979. Environmental Implications of Trends in Agriculture and Silviculture, Vol. 3. Regional Crop Production Trends. U.S. Environmental Protection Agency, Athens, GA. EPA-600/3-79-047.

Wauchope, R.D. 1978. The pesticide content of surface water draining from agricultural fields--a review. J. Environ. Qual. 7:459-472.

Wehtze, G.R., R.F. Spalding, O.C. Burnside, R. Lowry, and J.R.C. Leavitt. 1983. Biological significance and fate of atrazine under aquifer conditions. Weed Sci. 31:610-618.

Zake, M.H., D. Moran, and D. Harris. 1982. Pesticides in groundwater: The aldicarb story in Suffolk County, New York. Amer. J. Public Health. 72:1301-1395.

CHAPTER 20

AN EVALUATION OF THE APPLICABILITY OF SELECTED MODELS
TO CONSERVATION TILLAGE IN THE NORTH CENTRAL STATES

David B. Beasley
Department of Agricultural Engineering
Purdue University
West Lafayette, Indiana   47907

INTRODUCTION

There is a commonly held belief that "truth" exists in the sample bottle. In other words, that the only sure way to assess a water quality problem or the impact(s) of a change in management is to collect a number of bottles of water.

There is little doubt that we can very accurately measure the contents of each sample bottle. However, establishing "cause and effect" from these samples is difficult, if not impossible. Only if the sampled area is uniform in land use, topography, soil type, and management can we make sweeping comments about yields from specific situations. Year to year cropping, management, or weather variations can all create large differences in a watershed's hydrologic and erosion responses. Thus, determining trends or average yield values from short or even medium term monitoring programs is "iffy" at best and certainly fraught with problems. Even short term monitoring programs are expensive. Applicability of data gathered on one watershed to another watershed is questionable at best.

That is not to say that monitoring programs should not be carried out. Although monitoring is slow to produce results (which are very limited in "transportability"), the data is absolutely essential for determining if problems exist and what the absolute levels of yield are for certain events. Monitoring information is also essential for model development and validation. Deterministic or stochastic modeling approaches require physical information for optimization or calibration of modeling parameters.

While there are many models that are advertised as being planning models, some are definitely more suited to this task than others. Depending on the needed information and the resources for providing data to the model, there may be one or more "right" models. The model may be event-oriented or continuous, field, watershed, regional or national scale, distributed or lumped. In any case, the user should determine the strengths and weaknesses of potential planning models and match them to the available data and other resources such as time, money, computer, etc.

A DISCUSSION OF MODELING APPROACHES

Three different classifications of models will be described in this paper. Table 1 lists the classifications, some characteristics of models in each grouping, and at least one "working" model that is representative of the group. Obviously, a particular model will fit in each of the classification schemes. Thus, when the time comes to choose a model, the user will have to consider all three classes and determine which model(s) meet his criteria. Consideration of data required, monetary resources, and available computing power will also enter into the decision.

Scale of the Model

Models are used in the planning process for many different reasons. The scale of the model often but not always determines the level of descriptiveness of the simulation. A model developed for field scale or small watershed scale simulations generally can be expected to possess more descriptiveness than one designed for basin or regional planning. However, if only general information is required (estimated sediment yield reduction in a large area, for example), it is possible that some of the large scale models may be satifactory. On the other hand, if the planner requires information on the impact that site specific practice applications may have, then the smaller scale models should be utilized.

Another consideration that must enter into model selection is the degree to which a particular model describes the processes that are important in the planning program. If sediment yield is important and the model only predicts detachment (not transport or deposition), then that model

Table 1.   Model Classifications, Characteristics, and Examples.

Classification	Characteristics	Example Models
Field- or Watershed-scale	Detailed component relationships, continuous or event-oriented, lumped or distributed, more data intensive, some models produce very site specific information	CREAMS[a] ANSWERS[b] HSPF[c] SEDLAB[d]
Regional- or National-scale	Very general and less detailed relationships, typically continuous, generally lumped parameters, produces non-site specific information	WATERSHED[e]
Event-oriented	Generally more descriptive hydrologic relationships, typically requires more input data, usually produces more descriptive output	CREAMS ANSWERS SEDLAB
Continuous	Less descriptive hydologic relationships, may require long-term calibration data (hard to come by), less likely to produce the detailed data needed for site specific planning	HSPF WATERSHED CREAMS
Distributed Parameters	Superior for site specific simulations, generally event-oriented, usually watershed- or smaller scale, can be used to produce graphic as well as tabular output	ANSWERS SEDLAB CREAMS
Lumped Parameters	Simplified (averaged) relationships, not as well suited for site specificity (unless on field scale), may be event or continuous simulation	CREAMS HSPF WATERSHED

[a]USDA, 1980
[b]Beasley and Huggins, 1982
[c]Johnason et al., 1980
[d]Borah et al., 1981
[e]Sonzogni et al, 1980

is not really appropriate for that application. If the level of resolution of a model is such that individual practices or fields cannot be described, then that model won't be satisfactory for studying farm scale planning.

Field or Watershed Scale. These models provide a planner with the capability of studying the impacts of management at the application level. Generally, the use of a smaller scale model requires a much higher degree of resolution in the input data and a corresponding increase in required computer resources.

This level of detail allows for describing the surface roughness, infiltration, and erosion resistance effects that various forms and application levels of conservation tillage can have. Any farm or watershed scale planning should be done with this type of model.

Regional or National Scale. These models are generally used for "screening" or gross impact analysis. They use long term, average data and linear relationships to produce estimates of gross sediment and/or nutrient yield (sometimes modified to net yields by "delivery ratios"). This level of detail is unacceptable for studying or adequately assessing conservation tillage impacts on water quality.

Time Frame of the Model

This classification is probably the most difficult to decide upon. Short term or event-oriented simulations provide data and understanding on watershed responses when the system is most active. Continuous or long term simulations provide information on average annual yields or yields to be expected from specific patterns of weather.

Since the majority of sediment yield and particulate-related chemical yields will occur as a result of the few large events in a year, event-oriented models can be used very effectively for these situations. However, if soluble chemicals or water yield is the "target" of the simulation, a continuous model is probably more appropriate.

Event-Oriented Simulation. The time increment between calculations is very small in these models. They are designed to produce very accurage simulations of the timing, shape, and magnitudes of the hydrographs and other storm-related phenomena. These models are typically more spatially descriptive than longer term models.

Continuous Simulation. These models use longer time increments and "coarser" descriptions of precipitation (i.e., daily or monthly totals). They produce information on long term yields and utilize actual or effective rainfall patterns as input. The hydrograph is often produced using a shape function or a calibrated, regression-type of relationship. The erosion and hydrology components are, by necessity, less descriptive than in a model with a shorter time increment.

Spatial Descriptiveness in the Model

There are many different opinions as to the level of detail needed for adequate planning. The degree of spatial and physical descriptiveness in a model has a dramatic impact on its applicability in specific planning situations. The use of empirical versus theoretical or calibrated versus deterministic modeling techniques also can have dramatic impacts on modeling costs and results.

Lumped parameters often simplify the input data and component relationships in the model. However, these simplifications may undermine the ability of the user to study the impacts of specific placements of tillage or structural practices. Distributed parameters provide much more spatial and physical descriptive potential. They also, by definition, require a more complex modeling framework and greater computer and data resources.

The determining factor often comes down to resource limitations. Limited money or personnel may shift the balance toward lumped parameters. On the other hand, when site specific placement data or uncalibrated modeling capabilities are needed, distributed parameters may be required. In general, a balance must be struck between available resources and the required information.

Lumped Parameters. A lumped parameter model, by definition, assumes that the overall system response can be represented by several "averaged" parameters and essentially linear relationships. One of the major problems with this type of model is that the modeler (or model builder) must basically know what the response will be to a given input before it is applied. The reason for this is that the model is not truly representing the physical system but, instead, a group of parameters that can produce responses that are similar to the actual system.

Lumped models often require calibration. This means that the model's internal parameters are adjusted to produce outputs consistent with observed information. Although there are deterministic equations describing the component relationships, they generally don't have a lot of physical significance.

While lumping typically simplifies the model and reduces computer requirements, it also reduces the model's ability to describe spatial placement of land uses and management practices. Thus, when comparing alternative management scenarios, a model requiring calibration and using lumped parameters may not be applicable.

Distributed Parameters. A distributed parameter model attempts to describe the spatial configuration of the watershed. Just as with lumped parameter models, there are varying degrees of distributed parameter models.

While most distributed parameter models do not require calibration for relationship adjustment, they should be verified and, if needed, optimized for the watershed(s) they are being applied to. The verification and optimization process entails simulating observed events with "assumed" values of pertinent system parameters. Most of these values will have been determined through moisture balance equations, field surveys, or from soil surveys. Generally, most of the parameters are actually presented as a range of numbers and can be legitimately modified to meet the specific conditions of the storm and watershed in question.

A direct consequence of distributed parameters is a more complicated modeling structure and a requirement for more detailed input data. This generally translates to a more expensive modeling system. However, the output should also be more detailed and descriptive of the specific watershed configuration and antecedent conditions.

Although distributed parameter models aren't necessarily more deterministic, they generally use more physically descriptive relationships. This, coupled with their physical configuration, allows for a priori description of the pertinent system parameters and negates the need for long calibration data sets.

SUMMARY

Three different classification schemes for models have been presented. Any model that is to be used for conservation tillage or any other soil erosion or water quality application will fit into each of these three schemes.

The potential user can determine the characteristics (or constraints) that apply to his particular application by using this classification system. Several specific models are presented and their characteristics classified through the use of the system presented.

LITERATURE CITED

Beasley, D.B., and L.F. Huggins. 1982. ANSWERS User Manual. EPA-905/9-82-001. U.S. Environmental Protection Agency, Region V. Chicago, IL.

Borah, D.K., C.V. Alonso, and S.N. Prasad. 1981. Single event numerical model for routing water and sediment on small catchments. Appendix I, "Stream Channel Stability." A report by the USDA Sedimentation Laboratory, Oxford, MS, to the U.S. Army Corps of Engineers, Vicksburg District, Vicksburg, MS.

Johanson, R.C., J.C. Imhoff, and H.H. Davis. 1980. Users Manual for Hydrological Simulation Program -- FORTRAN (HSPF). EPA-600/9-80-015. U.S. Environmental Protection Agency, Athens Environmental Research Lab. Athens, GA.

Sonzogni, W.C. T.J. Monteith, T.M. Heidtke, and R.A.C. Sullivan. 1980. WATERSHED -- A Management Technique for Choosing Among Point and Nonpoint Control Strategies. Part I -- Theory and Process Framework. Great Lakes Basin Commission. Ann Arbor, MI.

USDA. 1980. CREAMS: A Field-scale Model for Chemicals, Runoff, and Erosion from Agricultural Management Systems. W.J. Knisel, ed. Conservation Report No. 26. Science and Education Administration, USDA.

CHAPTER 21

SOIL AND NUTRIENT RUN-OFF LOSSES WITH CONSERVATION TILLAGE

Boyd G. Ellis
Department of Crop and Soil Sciences
Michigan State University
East Lansing, Michigan 48824

Arthur J. Gold
University of Rhode Island
Kingston, Rhode Island 02881

Ted L. Loudon
Department of Agricultural Engineering
Michigan State University
East Lansing, Michigan 48824

INTRODUCTION

Conservation tillage has been defined by the Soil Science Society of America as "any tillage sequence which reduces loss of soil or water relative to conventional tillage." And conventional tillage is defined as "the combined primary and secondary tillage operations normally performed in preparing a seed-bed for a given crop grown in a given geographical area."[1] It is clear that a wide range of tillage conditions could be discussed under this topic. Equally important is the realization that there will be a strong interaction between soil properties, tillage operations and soil and nutrient run-off losses.

The introduction of conservation tillage in most forms has changed a number of factors important to soil

----

[1]Glossary of Soil Science Terms. October, 1979. Soil Sci. Soc. Am.

and nutrient loss. It must be remembered that the method of fertilizer application is normally changed with the tillage system.· If a moldboard plow is utilized in a conventional tillage operation, the fertilizer applied will be mixed with the plow layer at least once a year. But many types of conservation tillage do not mix the soil thoroughly, and the fertilizer is placed either on the surface of the soil or in a band just under the soil surface. Effects from this are of necessity long-term in nature. Yet many of the research studies to date have included too few years to provide definitive results.

In the last decade conservation tillage practices have gained support in the Great Lakes Basin because they do curtail sediment loss. Conservation tillage systems rely on surface crop residues to reduce surface water runoff and soil losses. The extent of surface cover varies with the crop grown prior to tillage and the implement used. The surface residue associated with conservation tillage practices ranges from no-till, where planting occurs in the undisturbed residue of the previous crop, to modified tillage practices such as chiseling and disking, which leave residue on 20 to 80 percent of the ground surface.

Surface residue limits sediment loss by reducing soil detachment and transport. During a high intensity storm residue will intercept a portion of the rainfall, diminishing the raindrop energy impacting the soil surface and reducing both soil detachment and surface sealing. Conservation tillage has been found to limit runoff velocities and sediment transport due to the increase in the hydraulic roughness of the field (Romkens et al., 1973; Niebling and Foster, 1977).

The frequency and timing of high intensity storms influence the effectiveness of conservation tillage. Conservation tillage is most effective at reducing nutrient and sediment losses in regions where overland runoff results from high intensity storms during periods without a crop canopy. Compared to the rest of the eastern U.S., the Great Lakes Basin experiences fewer high intensity storm events (Wischmeier and Smith, 1978). Within the basin the frequency of erosive storms declines in a northward direction.

We will primarily address two field monitoring studies undertaken in the northern agricultural lands of the basin to evaluate conservation tillage as a pollution abatement practice on sites that do not regularly experience high intensity storms. First, tillage effects on phosphorus, nitrogen and sediment losses from tiled croplands with very little slope and with relatively fine textured soil are presented. The first study was conducted to help define pollution losses from the watershed of Saginaw Bay in Lake Huron, where agriculture constitutes 41 percent of the entire watershed land use (Limnotech, 1983). The second area discussed will be non-tiled watersheds with relatively coarse textured soils. These will generally include soils with greater than 6 percent slope that are susceptible to considerable erosion.

Figure 1 shows the approximate locations of the two study sites. Site 1 is a pair of watersheds which are the basis for the study of the coarse textured, more sloping soil area. Site 2 is the flatter, fine textured soil area.

FINE-TEXTURED SOILS WITH LITTLE SLOPE

The basis for this part of the paper is a pair of approximately 4 ha fields located in the southeast Saginaw Bay drainage basin (Site 2). Natural hydrological events were monitored for over two years. Figure 2 is a schematic diagram of the of the research site.

Figure 1.   Location of study sites; Site 1 is coarse-textured sloping watersheds and Site 2 is low slope fields with fine-textured soil.

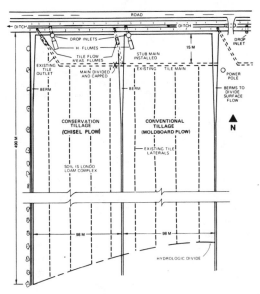

Figure 2. Skematic diagram of fine-textured field site with slopes less than 1 percent.

    The two fields studied were nearly identical with regard to slope, soil type and drainage patterns. With exception of primary tillage, the same agronomic practices were performed on each field. Slopes were approximately 1 percent. The soil, known locally as Londo loam, was a somewhat poorly drained alfisol, classified as an aeric glossaqualf, fine loamy mixed mesic soil. The primary tillage utilized on the conservation tilled field was a chisel plow which left 50 to 60 percent of the soil surface covered with corn residue. A moldboard plow which buried all crop residue was the primary tillage method on the conventional field.

    Both the overland and subsurface tile flow from each field were monitored. Most croplands in the study region have improved drainage as a result of subsurface tile. The study plots selected had a complete subsurface tile drainage network, and the subsurface flow was measured with a 60° triangular critical depth flume. Water samples for chemical analysis were obtained just upstream from the measurement flume using an ISCO model 1680 sampler. The overland flow on each field plot naturally drained to a single point where it was measured and sampled as it flowed through an H-flume and was discharged to a receiving ditch through a drop spillway.

Table 1.   Loss of water, sediment, phosphorus and nitrogen from conventional and conservation tillage on flat, tiled fields for a two year period.

Parameter	Period	Conventional Runoff	Conventional Tile Flow	Conservation Runoff	Conservation Tile Flow
		– – – – – – cm – – – – – –			
Water	Total	19.08	25.47	11.40	24.17
	18 events	18.54	24.41	11.40	23.31
	6/20/82	0.54	1.06	0.00	0.86
		– – – – – kg/ha – – – – – –			
Sediment	Total	771	220	233	140
	18 events	507	206	233	133
	6/20/82	264	14	0	7
Soluble P	Total*	0.30	0.20	0.17	0.23
	18 events	0.268	0.19	0.17	0.22
	6/20/82	0.032	0.007	0.00	0.01
Total P	Total	0.96	0.42	0.48	0.45
	18 events	0.66	0.38	0.48	0.42
	6/20/82	0.30	0.04	0.00	0.03
Nitrate	Total	5.96	35.0	4.8	21.7
	18 events	5.89	33.8	4.8	20.9
	6/20/82	0.07	1.2	0.0	0.8
TKN	Total	5.02	1.89	2.54	1.97
	18 events	4.26	1.82	2.54	1.92
	6/20/82	0.76	0.07	0.00	0.05

*Estimated from data for approximately 65% of events.

Results of two full years of field monitoring spanning from April, 1981 to April, 1983, are reported here. There were 19 hydrologic events including three from snowmelt and 16 resulting from rainfall. Table 1 lists the total losses observed over the 24 month period from each field from both surface runoff and tile flow. The crops grown during that time included corn in 1980 and 1981, and beans in 1982.

The conventionally tilled field lost substantially more sediment, as well as more total phosphorus and nitrogen, than the conservation tillage field. Sediment loss from both fields was low; snowmelt runoff generated the largest quantity of sediment from each field. A single intense storm that occurred in June, 1982, on an emerging bean crop accounted for much of the difference in sediment and phosphorus loss from the two tillage systems. This was the only occurrence in three years of a high intensity storm that generated overland flow when no crop canopy covered the soil surface. The probability of a runoff producing storm at the study location at this time of year is 0.24 (Gold et al., 1984).

Along with reducing sediment losses, conservation tillage can alter the hydrology of a well-drained site by improving the infiltration capacity of the soil. The combined volume of overland and tile flow from the two fields was slightly more for the conventional tillage field. The conservation tillage field had significantly more water infiltrate and exit via subsurface tile than the conventionally tilled field. Subsurface tile flow on both fields had significantly lower concentrations of phosphorus and sediment than overland flow (Table 2). But nitrate nitrogen concentrations were higher in the tile flow.

The combination of tile drainage with conservation tillage on these flat, heavy soils seems to be an important aspect of achieving water quality improvement. Conservation tillage increases infiltration but only if the soil is no longer saturated from the previous rainfall event. Subsurface tile will affect the hydrology of a site, reducing the magnitude and frequency of overland runoff. Following a rainfall event, artificially drained sites will rapidly reach field capacity, increasing the infiltration rate and moisture storage capability of the soil.

The tile drainage systems in the fields at Site 2 were capable of removing most of the gravitational water in the soil profile within four days of a rainfall or snowmelt event. During the six weeks following snowmelt in both years of monitoring, no overland flow was generated although several precipitation events of approximately 25 mm magnitude occurred.

Table 2.   Concentration of sediment, phosphorus and nitrogen
in runoff and tile flow for conventional and
conservation tillage in flat fields.

| | | Conventional | | Conservation | |
		Runoff	Tile Flow	Runoff	Tile Flow
		- - - - - - mg/l - - - - -			
Sediment	All events	404	61	204	51
	6/20/82	4890	94	-	59
Soluble P	All events	0.28	0.07	0.21	0.12
	6/20/82	0.59	0.05	-	0.09
Total P	All events	0.50	0.12	0.42	0.17
	6/20/82	5.55	0.28	-	0.35
Nitrate	All events	3.1	9.8	4.2	7.9
	6/20/82	1.29	7.9	-	8.1
TKN	All events	2.63	0.53	2.23	0.72
	6/20/82	14.1	0.47	-	0.47

As water percolates to subsurface drainage tiles,
most of the sediment transported by surface water is filtered
out by the soil medium.   Schwab et al. (1980), conducted
a long-term field investigation in the southern portion
of the Lake Erie Basin to study sediment loss from tiled
and untiled plots.   Mean flow weighted concentrations
of sediment in the subsurface drainage water were 50 to
90 percent lower than in the surface flow.   Skaggs et
al. (1982) used a simulation model to investigate the
influence of tile drainage on sediment loss.   For the
sites modeled, yearly sediment losses were calculated
to be 9000 kg/ha for the untiled conditions and 900 kg/ha
when subsurface tile was present, a 90 percent reduction
in sediment leaving agricultural land.

In a study of subsurface agricultural drainage water
in the drainage basins of Lake Michigan and Lake Huron,
Erickson and Ellis (1971) concluded that tile drainage

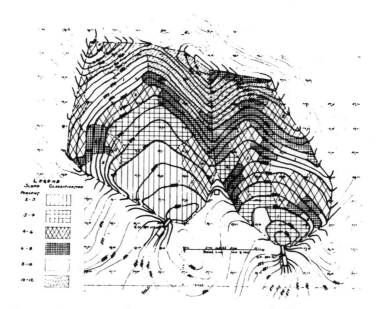

Figure 3.   Topographical survey of runoff plots.   Plot
06 is on the left and 07 is on the right.

carried low concentrations of soluble phosphorus.   The
flow weighted phosphorus concentrations in the subsurface
tile drainage from study Site 2 of our own investigations
were higher for the conservation tillage than for conventional
tillage but much less than the flow weighted phosphorus
concentration found in the overland flow from each plot
(Table 2).   The low concentrations observed in the tile
drainage water may have resulted from adsorption of soluble
phosphorus by the subsurface soil.   The study plots had
consistently received large amounts of phosphate fertilizer,
a common practice in highly productive agricultural areas,
and has more than 110 kg/ha of available phosphorus in
the surface layer (0-30 cm) of soil.   In contrast, the
subsurface soils had less than 10 kg/ha of available phosphorus
and could be expected to adsorb soluble phosphorus from
the water percolating to the subsurface tile.

The potential pollution abatement to Saginaw Bay
was calculated for different levels of conservation tillage
in the watershed (Limnotech, 1983).   Complete implementation
(100%) of conservation tillage was predicted to reduce
total net phosphorus loads to Saginaw Bay by 9.4 percent
and sediment loads by 44 percent compared to zero implementation
in the watershed.   Although these reductions are not as
great as might be expected in regions with more frequent

erosive storms, it was concluded that conservation tillage would be a cost effective method of pollution abatement.

COARSE-TEXTURED SOILS WITH APPRECIABLE SLOPE

A matched pair of approximately 1.2 ha (see Figure 3) watersheds have been studied for many years by Erickson and others at Michigan State University. During the period of 1972 through 1976 the watersheds were heavily fertilized with phosphorus (as well as other nutrients), farmed by conventional tillage practices and all run-off including pesticides, sediment and nutrient loss carefully measured. The results of this study have been published (Ellis et al., 1978 and Hubbard et al., 1982). A brief summary of the results are given below, and the interested reader may obtain complete results by referring to both publications.

Fifty-six run-off events occurred over the two year period of study (April, 1974 through March, 1976). However, in terms of sediment loss the large majority occurred during a single event on April 18, 1975. Summary results are given in Table 3. The two watersheds were rather similar in run-off loss and in sediment loss except for the single large event. Here the smaller watershed lost nearly 2.5 times as much sediment expressed on a unit basis (kg/ha). Phosphorus losses in the water phase were about one kg/ha-year. However, the sediment phase carried about 15 and 20 kg P/ha-year for watersheds 06 and 07, respectively. Nitrogen was a contrast to both sediment and phosphorus losses in that the majority of the nitrogen lost was not during the one single event in the water phase (i.e., about 35 percent of the phosphorus in the water phase came in the one event whereas about 18 percent of the nitrogen in the water phase came in the single event). Total soluble nitrogen lost was 12.5 and 13.1 kg/ha for watersheds 06 and 07, respectively. The nitrogen in the sediment phase was 52.4 and 59.7 kg/ha, respectively, for 06 and 07 with the majority coming from losses during the one event.

Summarizing these studies, it can be said that conventional tillage of watersheds with slopes greater than 6 percent lead to unacceptable losses of both soil and nutrients, particularly during major storm events.

Upon termination of the above studies, both watersheds were extremely high in phosphorus and susceptible to loss of phosphorus through both sediment and water run-off. Management of the watersheds was changed at that time to study two factors. First, could sediment and water

Table 3. Loss of water, sediment, phosphorus and nitrogen from two watersheds over a two year period (data from Hubbard et al., 1982).

Parameter	Period	Watershed Number	
		06	07
		- - - - - kg/ha - - - - -	
Sediment	Total	19,213	36,831
	56 Events*	6,981	5,165
	April 18, 1975	12,232	31,666
		- - - - - - cm - - - - -	
Water	Total	34.9	31.9
	56 Events*	26.8	23.3
	April 18, 1975	8.1	8.6
		- - - - - kg/ha - - - - -	
Phosphorus	Total	1.9	1.7
in Water	56 Events*	1.33	1.04
	April 18, 1975	0.57	0.66
		- - - - - kg/ha - - - - -	
Phosphorus	Total	29.6	39.6
in Sediment	56 Events*	13.3	8.71
	April 18, 1975	16.3	30.9
		- - - - - kg/ha - - - - -	
Nitrogen	Total	12.5	13.1
in Water	56 Events*	10.1	10.9
	April 18, 1975	2.4	2.2
		- - - - - kg/ha - - - - -	
Nitrogen	Total	52.4	59.7
in Sediment	56 Events*	21.0	14.3
	April 18, 1975	31.4	45.4

*This includes all events except for the major event of April 18, 1975.

runoff be controlled by a system of conservation tillage combined with intercropping; and second, what will the rate of phosphorus drawdown be if no phosphorus fertilizer is applied to either watershed.

Experimental Procedures

Watershed 06: This watershed was seeded to a small grain (rye and/or barley) and each year corn was intercropped by killing a strip of the small grain with paraquat in which the corn was planted with a no-till planter. Corn grain yields were estimated and the corn grain removed from this watershed. All other residue was left on the watershed. Soil samples were collected each year from six segments of each watershed and from the 0-1, 1-2.5, 2.5-5, 5-7.5, 7.5-15, and 15-30 cm depth of each segment by compositing 20 cores each 2.5 cm in diameter. Samples were air-dried, crushed and screened to pass a 2 mm sieve and extractable phosphorus (Bray-Kurtz P1) and water soluble phosphorus determined.

Watershed 07: This watershed was seeded to alfalfa and each year corn was intercropped by killing a strip of alfalfa with paraquat and planting corn in the strip with minimum tillage equipment. Corn grain yields were estimated and the corn grain removed but all other residue was left on the watershed. Soil samples were collected as on watershed 06.

The most noticeable effect of the change to a conservation management system was the decrease in runoff volume and sediment loss. No runoff has occurred during the growing season for eight years under the new intercropped systems. Some runoff has occurred during the winter and early spring months. This total runoff has contained negligible sediment but it has contained considerable soluble nutrients. This is shown in Table 4. Perhaps the most significant finding is that coupled with the great reduction in total runoff is an appreciable increase in the concentration of nutrients in runoff water, particularly of soluble phosphorus. This undoubtedly occurs because organic residues are accumulated on the surface under conservation tillage, and this leads to increased soluble nutrients in water that is in contact with this organic residue. But the net effect is to reduce total nutrient losses when conservation tillage is used.

Soil samples have shown a very interesting pattern during this experiment. This is illustrated in Figures 4 and 5. Initially, both watersheds were very high in phosphorus. By 1981, the watershed which was being cropped

with alfalfa and corn showed considerable reduction in
Bray-Kurtz P1. Values were still very high but they were
being reduced. However, the watershed which was being
cropped with rye and corn showed quite a different pattern.
The extractable phosphorus in the surface had not reduced
and in the depths of 7.5 to 30 cm it had, in fact, increased. The reason that this occurs is not very apparent.
It may be due to utilization of phosphorus by the rye
and corn from depths greater than 30 cm and redeposition
of this phosphorus near the surface where the plants die
at the end of each season. If this is the case, one would
have expected alfalfa to have done the same thing. But,
since it is a perennial and a tap root-type plant, the
phosphorus may still be largely retained in the living
plants. In any event, it is evident that soils high in
phosphorus are not reduced rapidly with cropping under

Table 4.   Concentration of nutrients in runoff from
watersheds 06 and 07 in 1980.

Date	06			07		
	SP	TP	N*	SP	TP	N*
- - - - - - - - - mg/liter - - - - - - - - -						
Conservation Tillage						
2/17/80	--	--	--	0.97	1.02	1.96
2/21/80	1.98	2.32	7.79	1.70	1.75	3.34
2/22/80	1.87	1.94	4.19	1.85	1.89	2.91
3/10/80	1.09	1.13	4.41	0.71	0.72	3.95
3/16/80	1.19	1.20	2.58	1.02	0.90	2.85
3/17/80	1.65	1.70	2.97	1.27	1.31	1.76
Conventional Tillage						
Avg 74-76	0.45	0.50	1.97	0.39	0.44	2.69

*Soluble ammonium and nitrate nitrogen expressed   as mg
N/liter.
SP = Soluble Phosphorus
TP = Total Phosphorus

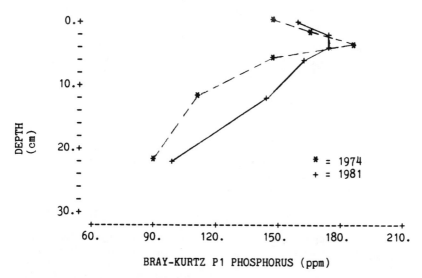

Figure 4. Bray-Kurtz P$_1$ phosphorus as affected by intercropping rye with corn on watershed 06.

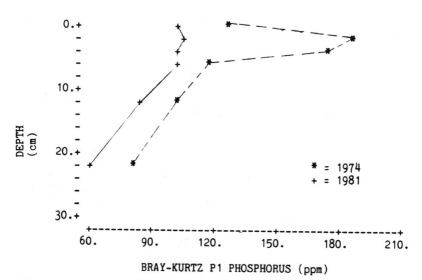

Figure 5. Bray-Kurtz P$_1$ phosphorus as affected by intercropping alfalfa with corn on watershed 07.

either of these conservation tillage systems. But conservation tillage practices may concentrate phosphorus in the surface of the soil resulting in increased phosphorus concentration in surface runoff waters.

The results of the above study are similar to others reported that are of a much shorter duration. Mueller et al. (1982), have studied the relationship between different conservation tillage methods and phosphorus runoff. The tillage treatments they were using were conventional (plow-disk), chisel plowed, till-plant and no-till. The data they gave was for simulated rainfall. During their first runoff period, runoff losses were reduced by 60 to 85 percent at conservation tillage sites relative to conventional tillage sites. Again, this difference is largely due to a reduction in total runoff, not to a reduction in concentration of nutrients in the runoff.

Other studies have shown increasing losses from conservation tillage methods. For example, Romkens et al. (1972) found higher concentrations of soluble phosphorus and sediment phosphorus with conservation tillage methods, particularly when a coulter was used as compared to conventional tillage. Even the total loss of soluble phosphorus was higher. We do not believe that this is generally true and may occur for a few short-term studies with certain conservation tillage methods. Further research will clarify this point.

SUMMARY AND OBSERVATIONS

Conservation tillage changes water flow patterns. Generally, more water penetrates the soil and may then emerge as tile flow in level tiled land or move to groundwater in porous non-tiled land. The following observations have been made about how this affects nutrient losses.

(1)   Conservation tillage reduces sediment loss, particularly for high intensity rainstorms.

(2)   Conservation tillage will reduce net total losses of sediment bound phosphorus and nitrogen although sediment may be enriched compared to conventional tillage.

(3)   Conservation tillage tends to increase concentrations of soluble phosphorus and nitrogen in runoff.

(4)   On sites with low erosion where much of phosphorus and nitrogen are lost in the soluble form,

conservation tillage may not be effective
at reducing losses of available and soluble
phosphorus.

(5)   Fertilizer application methods are very important
with conservation tillage.

A tabular summary of other studies which report
comparisons of water quality from conventional and conservation
tillage systems is included at the end of this paper as
Table A-1.

LITERATURE CITED

Ellis, B.G., A.E. Erickson, and A.R. Wolcott. 1978.
Nitrate and phosphorus run-off losses from small watersheds
in Great Lakes Basin. EPA-600/3-78-028. 84 p.

Erickson, A.E. and B.G. Ellis. 1971. The Nutrient Content
of Drainage Water from Agricultural Land. Res. Bul. 31.
Michigan Agricultural Experiment Station.

Gold, A.J., T.L. Loudon, and F.V. Nurnberger. 1984.
Long term weather records to assess best management practices.
ASAE Paper No. 84-2043, Am. Soc. of Ag. Engrs., St. Joseph,
MI.

Hubbard, R.K., A.E. Erickson, B.G. Ellis, and A.R. Wolcott.
1982. Movement of diffuse source pollutants in small
agricultural watersheds of the Great Lakes Basin. J. of
Environ. Qual. 11:117-123.

Limnotech, 1983. Cost Effectiveness of Conservation Tillage
Practices for Reducing Pollutant Loads to Inner Saginaw
Bay. ECMPDR. Saginaw, MI.

Mueller, D.H., T.C. Daniel, B. Lowery, and B. Adraski.
1982. The effect of conservation tillage on the quality
of run-off water. Paper for Amer. Soc. of Ag.Eng.

Neibling, W.H. and G.R. Foster. 1977. Estimating Deposition
and Sediment Yield from Overland Flow Processes. In:
Proceedings International Symposium on Urban Hydrology,
Hydraulics and Sediment Control. University of Kentucky,
Lexington, KY, pp. 75-86.

Romkens, M.J.M., D.W. Nelson, and J.V. Mannering. 1973.
Nitrogen and phosphorus composition of surface run-off
as affected by tillage method. J. Environ. Qual. 2:292-295.

Schwab, G.O., N.R. Fausey, and D.E. Kopcak. 1980. Sediment and Chemical Content of Agricultural Drainage Water. Transactions of the ASAE. Vol 23. SW 1980, pp. 14465-1449.

Skaggs, R.W., A. Nassehzadeh-Tabrizi, and G.R. Foster. 1982. Subsurface Drainage Effects on Erosion. Journal of Soil and Water Conservation. Vol. 37, No. 3, pp. 167-172.

Wischmeier, W.H. and D.D. Smith. 1978. Predicting rainfall erosion losses - a guide to conservation planning. USDA Agricultural Handbook No. 537. Washington, D.C.

LITERATURE CITED IN TABLES A-1 AND A-2

Baker, J.L. and J.M. Laflen. 1982. Effect of corn residue and fertilizer management on soluble nutrient losses. Trans. ASAE. 25:344-348.

Barisas, S.G., J.L. Baker, H.P. Johnson, and J.M. Laflen. 1978. Effect of tillage systems on runoff losses of nutrients, a rainfall simulation study. Trans. ASAE. 21:893-897.

Harrold, L.L., G.B. Triplett, Jr., and R.E. Yonker. 1967. Watershed tests of no-tillage corn. Journ. of Soil and Water Cons. 22:98-100.

Holt, R.F., M.P. Johnson, and L.L. McDowell. 1973. Surface water quality. In: Conservation tillage: The proceedings of a national conference. SCSA. pp. 141-156.

Johnson, C.B. and W.C. Moldenhauer. 1979. Effect of chisel versus moldboard plowing on soil erosion by water. SSSA Journal. 43:177-179.

Johnson, H.P., J.L. Baker, W.D. Shrader, and J.M. Laflen. 1979. Tillage systems effects on sediment and nutrients in runoff from small watersheds. Trans. ASAE. 22:1106-1114.

Laflen, J.M., J.L. Baker, R.O. Hartwig, W.F. Buchele, and H.P. Johnson. 1978. Effect of residue on runoff and erosion. Journ. of Soil and Water Cons. 33:107-109.

Laflen, J.M. and T.S. Colvin. 1981. Effect of crop residue on soil loss from continuous corn cropping. Trans. ASAE. 24:605-609.

Mannering, J.V. and L.D. Meyer. 1963. The effects of various rates of surface mulch on infiltration and erosion. SSSA Proc. 30:101-105.

McDowell, L.L. and K.C. McGregor. 1980. Nitrogen and phosphorus losses in runoff from no till soybeans. Trans. ASAE. 23:643-648.

McGregor, K.C. and J.D. Greer. 1982. Erosion control with no till and reduced till corn for silage and grain. Trans. ASAE. 25:154-157.

Romkens, M.J.M., D.W. Nelson, and J.V. Mannering. 1973. Nitrogen and phosphorus composition as affected by tillage method. J. of Environ. Qual. 2(2):292-295.

Siemens, J.C. and W.R. Oschwald. 1978. Corn-soybeans tillage systems: erosion control, effects on crop production costs. Trans. ASAE. 21:293-302.

Smith, G.E., F.D. Whitaker, and H.G. Heineman. 1974. Losses of fertilizers and pesticides from clay pan soils. Off. Res. and Devel. U.S. EPA. technology series. 6602-74-068.

Wischmeier, W.H. 1973. Conservation tillage to control water erosion. In: Conservation tillage: The proceedings of a national conference. SCSA. Ankeny, Iowa. pp. 133-141.

Table A-1.   Summary of Literature Involving Rainfall Simulator
Plot Studies Comparing Sediment and Nutrient
Loss for Conservation vs. Conventional Tillage.

SOIL	TILLAGE TYPES	SEDIMENT	SOLUBLE PHOSPHORUS
	Artificial residue placement—0 + 3 levels; fert. vs. nonfert. under and over residue; also clean tilled 36 plots	Decrease with residue	Phosphate conc. higher with increased residue
3 soils	6 prac.—CVT, till-plant, CH, disk, ridge, NT 36 plots	Sediment was main carrier of nutrients	Increased residue increased avail. P conc. on sediment. More loss from CH and till-plant. Increased residue caused increased soluble nitrate conc.
	Fall MB, fall CH, sp. disk; 1st no residue; then all covered with 20% residue	CH alone decreased soil loss by 75%; with residue CH had 50% less soil loss	
3 soils	Same as Barisas, et al., 1978 above	Increased residue reduced sediment conc. in runoff. Corn residue more effective than straw	
2 soils	MB, CH, and NT; 3 canopy cover levels 10%, 60% abd 85%	10% canopy sign. diff. NT<CH<MB; 60% canopy NT<CH<MB; 85% canopy NT<MB; 2 soils diff. residue very important	

SEDIMENT PHOSPHORUS	NITROGEN	COMMENTS	AUTHOR(S)
		Not truly representative of CST. Residue on unfertilized plots did not increase nutrient conc. significantly	Baker and Laflen, (1982)
Only disk, ridge and NT reduces P loss. More P loss from CH and till-plant than CVT	CST reduced N loss by reducing erosion. Residue had no effect on N conc.		Barisas, et al., (1978)
		CH leaves residue and affects surface condition, decreasing erosion	Johnson and Moldenhauer, (1979)
		Soil loss highly related to residue cover with no crop canopy	Laflen, et al., (1978)
			Laflen and Colvin, (1981)

Table A-1.   Summary of Literature Involving Rainfall Simulator Plot Studies Comparing Sediment and Nutrient Loss for Conservation vs. Conventional Tillage. (Continued)

SOIL	TILLAGE TYPES	SEDIMENT	SOLUBLE PHOSPHORUS
5% slope	Straw cover, 6 levels from 0 to 100%	Increased residue decreased erosion	
6% slope plano silt loam	MB, CH, till-plant, NT	NT reduced sediment to 3% of MB	Reported and other forms
8-12% slope silt loam	MB, CH (on contour), buffalo till, disk, NT	Loss % of MB; CH 5%, buffalo 37%, disk 15%, NT 7%	Conc. increased as soil loss decreased
	7 tillage systems; spring meas. before canopy	Soil loss more for fall plow than CST	Did not vary sign. between treatments
	NT, CH, MB; corn--2 crop stages	NT reduces soil loss 82-95%; CH up and dn slope reduces loss 48% at early growth stage (4 wks) and 30% @ 35% canopy compared to MB	

CST = Conservation Tillage   CH = Chisel Tillage
CVT = Conventional Tillage   P = Phosphorus
MB = Moldboard Tillage   N = Nitrogen
NT = No Tillage

SEDIMENT PHOSPHORUS	NITROGEN	COMMENTS	AUTHOR(S)
			Mannering and Meyer, (1963)
			Mueller, et al., (1982)
MB loss > CST; conc. on sediment higher for CST than MB	MB loss > CST; conc. on sediment higher for CST than MB; conc. of soluble N increases as soil loss decreases	CST controls erosion but may not always limit nutrient loss	Romkens, et al., (1973)
Greater loss with fall plow than all CST methods	Nitrate did not vary sign.; greater sedimentation associated N loss with fall plow than CST	Less water run-off with CH	Siemans and Oschwald, (1978)
			Wischmeier, (1973)

Table A-2.   Summary of Literature Involving Field Studies of Natural Runoff Comparing Sediment and Nutrient Loss for Conservation vs. Conventional Tillage

SOIL	TILLAGE TYPES	SEDIMENT	SOLUBLE PHOSPHORUS
9.4% slope	MB and NT	3 yr study total loss: MB = 6.7 MT/ha NT = 0.1 MT/ha	
Steep erodible soils	MB, disk-plant ridge plant (NT) 6 watersheds of 0.55-1.75 ha 3 seasons sp tillage	Soil loss for 3 seasons: MB = 25.4 MT/ha D-P = 12.2 MT/ha NT = 1.8 MT/ha	Conc. higher for CST; also more total loss for CST
5% slope silt	MB and NT (0.01 ha plots)	NT loss 0.7 – 12.2% of MB	Conc. higher for NT. Conc. higher in May and June (after planting) for all plots
5% slope silt loam	MB and NT	3 yr study NT soil loss 4% of MB NT<0.5 MT/ha/yr	
3% slope clay-pan	MB and NT C and SB small plots	3 yr avg annual loss: MB = 1.46 MT/ha NT = 0.22 MT/ha	NT had greater total P loss than MB

CST = Conservation Tillage
CVT = Conventional Tillage
MB = Moldboard Tillage
NT = No Tillage
CH = Chisel Tillage
MT = Metric Ton

SB = Soybeans
W = Wheat
C = Corn
P = Phosphorus
N = Nitrogen

SEDIMENT PHOSPHORUS	NITROGEN	COMMENTS	AUTHOR(S)
			Harold, et al., (1967)
Avail. P conc. on sediments was greater for CST	More net loss from MB site. Con. of ammonia and nitrate not diff. for CST most N loss was sediment bound	CST reduced run off and sediment loss but high soluble P conc. resulted in more soluble P loss for CST	Johnson, et al., (1979)
Sediment bound conc. of N and P were 5 fold more for NT than MB. Total P loss reduced 41-87% by NT	Conc. of soluble N higher for NT Total - N loss reduced 49-87% by NT	Rotations went SB-SB, SB-W, SB-C, C-SB	McDowell and McGregor, (1980)
			McGregor and Greer, (1982)
Less than soluble P (note low sediment loss)	No diff. in nitrate and ammonia loss		Smith, et al., (1974)

CHAPTER 22
CONSERVATION VS. CONVENTIONAL TILLAGE:
ECOLOGICAL AND ENVIRONMENTAL CONSIDERATIONS

Maureen Kuwano Hinkle
National Audubon Society
Washington, D.C. 20003

INTRODUCTION

Developed for military uses in World War II, pesticides became a major factor in pest control programs. Chemical control of insects meshed well with the post-war economic, social and technological advances and rapidly became integrated in the U.S. food-production system. The dominance of chemical control then became inevitable because of its neat structural fit in the farmer's battle to stay in business as two-thirds of farmers were forced off the farm into other businesses. By reducing the risk of crop losses to pests, farmers could increase credit for enlarging their operation. Reliable insect control became security for increased crop production, rendered increasingly more vulnerable by the very chemical weapons themselves. Insect resistance and resurgence of secondary pests as well as problems of human health, wildlife and the environment developed.

In response to these and other environmental problems, the Environmental Protection Agency (EPA) was created in 1970 to regulate toxic substances. In 1972, the Federal Insecticide, Fungicide and Rodenticide Act (FIFRA) was amended to give EPA a comprehensive mandate to regulate pesticides. Pesticide regulation, however, proved to be elusive, politically hypersensitive, and crisis-dominated—a roller coaster rather than an orderly process throughout the decade.

Despite this new authority and increased public awareness, pesticides increased in volume of use and sales

until 1981.   By 1981 the chemical pesticides industry
matured independent of regulatory impact on its growth.
In response to a slowing of anticipated expansion and
growth potential, the chemical industry is seeking new
markets, one of which is conservation tillage.

The dynamics of pesticide use changed dramatically
in the 1970s. Insecticides had been the largest percentage
of pesticide use in the United States in the 1960s and
early 1970s. Today, they constitute less than 25 percent
of pesticide use in the United States. Herbicides, on
the other hand, increased 73 percent between 1966-77,
and now constitute approximately 67 percent of all pesticides
used in the United States. Of exports, insecticides continue
to comprise the larger percentage in volume and sales,
currently accounting for one-third of all pesticides produced,
and projected to increase fourfold in the next 20 years
in developing countries.

The promise of conservation tillage is extraordinary
as a means to reduce soil erosion, halt sedimentation,
reduce energy and labor, increase water efficiency, and
reduce runoff of chemicals. It is also less costly as
a soil conservation measure than structural measures such
as contouring, terracing, and diversions. Congress, USDA,
and others operating in an agricultural economy dominated
by the production of cash grains hope to cut production
costs and conserve soil with wide-scale adoption of conservation
tillage. Some projections for conservation tillage have
been as high as 100 percent of croplands by the year 2000.

Adoption of conservation tillage, however, often
involves use of more herbicides and, after several years
of continuous conservation tillage, increased use of other
pesticides as well. Compared with conventional tillage,
the increase in herbicides ranges from a 1984 survey by
the Conservation Tillage Information Center (19%) to Purdue's
Griffiths and Parsons in 1983 (43%). The cost of removal
of recently detected herbicides from municipal water supplies
is yet another factor that has not been researched, explored,
or economically factored. This paper discusses the added
dimensions conservation tillage brings to environmental
and ecological concerns.

ENVIRONMENTAL CONCERNS OF PESTICIDES USED ON CROPLANDS

Large quantities of agricultural chemicals are currently
used in crop production regardless of the type of tillage
system. In 1982, the most recent year for which figures
are available, United States industries produced 682 million
kg of pesticides, excluding wood preservatives which account

for an additional 455 million kg. Approximately 1,400 active ingredients are formulated into 50,000 registrations on the market today.

Environmental problems have resulted from application of pesticides to croplands both on-site and off-site. On-site, insects, fungi, nematodes, weeds and other pests have resurged and become resistant to chemical pesticides used in greater quantities in efforts to control them. Pesticides have persisted in target areas and adversely affected nontarget species inhabiting the same ecosystem. Off-site, agricultural chemicals have been carried far distant from the target areas and have not only disrupted adjacent and even distant ecosystems, they have also destroyed sensitive nontarget crops. 2,4-D, for example, was applied to wheat in the Northwest and destroyed valuable grape crops.

An understanding of environmental fate of pesticides remains inadequate. How pesticides move through and are transformed in the environment is not well understood. They can be transported to surface waters in runoff, leached through soil to groundwater, and volatilized from the soil or foliar surface to the atmosphere. Some pesticides bind to soil particles, move in vapor stage, or are transported in rain. Some degrade into substances more persistant than the parent compound.

Agriculture appears to be in transition, with conservation tillage rapidly increasing, particularly onto pasture and erosion prone lands otherwise not suitable for cultivation.[1] Potentially more as well as different kinds of pesticides may be required for such lands.

The shift from conventional to reduced tillage systems is accompanied by a corresponding increase in soluble pesticides and also a return to persistent type pesticides which are more likely to move to groundwater and are probably even more persistent in the subsoil. Research on persistence in the subsoil as well as movement to groundwater is needed.

Herbicides

Because standard herbicide treatments may be less effective due to residue tie-up on the soil surface, more

---

[1]Conservation tillage doubled between 1973-81 (total increase was 49 million acres), now practiced on 30 percent of croplands.

herbicides may be needed with reduced tillage. Moving from conventional to no-till corn, herbicide application has been observed to increase from a total of 3.89 to 5.60 kg/ha, a 43 percent increase. Chisel plowing requires 4.49 kg/ha, a 14 percent increase over conventional tillage (Griffiths and Parson, cited in Christensen and Norris, 1983). Projections for increased use of pesticides for no-till corn are as high as 50 percent compared to conventional tillage corn (Phillips, 1980, and USDA, 1975, cited Christensen and Norris, 1983). An increase of 40-50 percent when switching from conventional tillage to no-till has also been reported (Christensen and Norris, 1983).

In a recent study of cost estimates for a conventional plow system, an experimental two-till, a one-till, and zero-till system on an Oklahoma wheat farm, the data in Table 1 was developed.

Table 1. Estimates of labor, herbicide, fuel, and capital requirements and costs per hectare by tillage system (Epplin, et al., 1982).

	System				
	Plow	Two till	One till	Zero till	Plow/ One-till
Machinery labor (hours/ha)	3.09	0.99	0.96	0.62	2.49
Herbicide ($/ha)	3.14	28.01	37.57	52.02	20.35
Tractor fuel (liters/ha)	59.78	29.57	15.72	11.13	38.17
Annual operating capital ($/ha)	99.57	110.90	126.22	146.55	113.87
Average machinery investment ($/ha)	184.21	132.79	121.38	133.95	164.55
Operating + machinery capital ($/ha)	283.78	243.69	247.59	280.49	278.44
Total operating cost ($/ha)	207.95	215.63	225.51	253.57	220.08
Machinery fixed cost ($/ha)	55.62	41.17	37.35	40.76	50.71
Total operating + machinery cost ($/ha)	263.57	256.81	262.86	294.33	270.79

Machinery, labor, fuel and machinery investment was lower in reduced tillage systems. On the other hand, the requirement for more operating capital was more (11% to 47% annual), and herbicide costs higher (793% to 1,558%) than that estimated for the plow system. Costs for the zero-till system were 14% more than the plow system (Epplin, 1982).

One herbicidal use that has received little public notice is toxaphene, used for 20 years against sicklepod in the Southeast, but not officially registered as an herbicide until 1980. The National Cancer Institute and EPA concluded that toxaphene is an extremely potent laboratory carcinogen which also poses grave risks for fish and wildlife. In 1981 the estimated average number of acres with toxaphene as an herbicide came to nearly one million acres of soybeans (at 4.49 kg active ingredient/ha). Additional quantities are applied to soybeans and peanuts as an insecticide (EPA Economic Analysis, June, 1982). Its use on sicklepod in the Southeast (and Midwest on corn worms) will continue to distribute this potent chemical into the environment.

Insecticides

Pierre Crosson arrived at a rough 30 percent increase in pesticides with conservation tillage over conventional tillage systems (Crosson, 1981). His figure includes all pests, such as rodents, snails, worms, etc., not just insects. Exact figures are elusive as few multi-year studies have been conducted. Toxaphene is an example of the type of chemical sought for minimum till. In 1982 USDA asked EPA for continuation of the registration of toxaphene for use in small grains, soil insects and grasshoppers. By way of explanation, USDA stated that "soil insect populations are significantly higher in croplands under minimum tillage practices. Toxaphene is the most efficacious material for the control of insect populations located in the soil." (47 Federal Register, 53791, November 29, 1982). All other uses were cancelled.

Because of several crop-damaging insects, particularly in corn and soybeans, prophylactic use of insecticides is relied upon in reduced tillage systems without questioning the underlying problems. One such example is the recommendation to use 33.64 kg of carbofuran per ha of corn on a prophylactic basis.

Rodenticides

As planting directly into the stubble of the previous crop takes place under no-till, a suspected increase in populations of rodent pests has been reported in the meadow steppe, or Palouse prairie, of East-central Washington State and adjacent Northern Idaho (preliminary work by F.F. Gilbert, Director of Wildlife Biology, and B. Josephson, graduate student, at Washington State University, Pullman, 1983). Sporadic reports of increased rodent pests in no-tilled crops have circulated, but have not been described yet.

RESEARCH NEEDS--HERBICIDES

Although heavier herbicide use is expected, understanding of the implications of such use is limited. "Of the approximately 150 chemicals used as herbicides throughout the world, complete metabolic breakdown pathways are known for only three of four of these compounds. Metabolic degradation of the remaining compounds is known only in part" (Kaufman and Kearney, 1976). To correct this paucity of vital information, "[W]hat is clearly needed is an administrative protocol for assessing the significance of new metabolites not reported at time of registration" [emphasis added]. This can be accomplished with "...a continuing sophisticated approach to identifying and quantifying the major and minor herbicide metabolites from all biological systems. Only when these are known can their significance be accurately gauged" (Kaufman and Kearney, 1976).

As J.M. Way and R.J. Chancellor (1976) put it, because "...herbicides represent a collection of chemical groups that all have high biologic activity over the plant kingdom, a greater understanding of their ecological consequences may in time become more important than the minutiae of their more detailed effects." The fate and transport characteristics of metabolites and/or transformation products of major pesticides are not known. Once they are known along with critical environmental parameters, simulation models can be used to evaluate their fate and transport properties.

Because 455 million kg of herbicides are produced in the United States each year, and the proportion of herbicides is expected to increase, the need to understand the effects of herbicides on plant and animal growth, soil systems, water and human health is vital (McWhorter, 1978).

Whereas many chlorinated hydrocarbons bioaccumulated in man and caused liver tumors in rats and mice, several herbicides persist in ground and surface waters and cause a variety of tumors in laboratory animals. Alachlor, of which 45 million kg is applied each year primarily on corn and soybeans, causes tumors of the stomach, nose, lung, thyroid and causes ocular lesions[2] in laboratory test animals. Consuming low levels of carcinogenic herbicides in our drinking water poses additional risks for man.[3]

RESEARCH NEEDS--INSECTICIDES AND OTHER PESTICIDES

Studies are needed on insect community composition, seasonal dynamics, and ecosystem interactions in no-tillage and reduced tillage agroecosystems. At this time there are indications that beneficial effects may occur in certain no-till systems. Arthropod pest damage is well known. Less known are soil arthropods that can promote efficient nutrient cycling in agroecosystems. Already demonstrated in several forest ecosystems, "soil arthropods in effect substitute as millions of tiny 'plows,' breaking down particles of surface crop residue of no-till systems" (Personal communication with Garfield House, North Carolina State University, January 24, 1984). Researchers at North Carolina State University report that no-tillage systems commonly support a higher diversity of pest insects than conventional tillage systems (House and Stinner, 1983). Infestations of the lesser cornstalk borer were also deterred in no-tillage corn cropping systems. In Northern Georgia soybeans, the beneficial carabid beetle number and species diversity were often several times higher in no-tillage than in conventional tillage soybeans (House and All, cited in House and Stinner, 1983). Similarly, in Tennessee, no-till planting of soybeans behind wheat produced less severe septoria brown spot, anthracnose and cyst nematodes than in conventionally tilled soybeans. In all these cases, it should be pointed out that crop rotations were practiced.

Increases in nematodes have been attributed to tillage itself. "Anytime you move the soil, you move nematodes and spread them across the field," reports Melvin Newman, associate professor of plant pathology at the University

---

[2] EPA Registration Standard. November 20, 1984.

[3] Other herbicides indicating carcinogenic effects: Linuron--interstitial testicular adenomas in rats, liver adenomas in female mice; Norflurazon--liver neoplasms; Pendimethalin--uterine adenocarcinomas in rats; Sonalan--breast adenomas in rats; Daminozide, a growth regulator, causes a statistically significant increase in blood vessel tumors and lung tumors in both sexes of mice; kidney and liver tumors in male mice; colon tumors in hamsters and mice.

of Tennesee (Conservation Newsletter, August 8, 1983, "No-Till Holds Down Diseases, Nematodes," by Forrest Laws, associate editor).

"[N]o-tillage management creates a very different environment from conventional systems," according to House, "and are apparently more complex. Some pest species survive better in no-tillage systems, but so do a number of predators. Therefore, successful no-tillage management requires information from many components so that trade-off criteria can be established."

Because no-tillage and reduced tillage require more intensive management than conventional methods, House urges that strategies for pest management in no-tillage systems "focus not only on one crop, but rather on an entire crop management system, including rotations and cover cropping" (House and Stinner, 1983).

According to Rodriquez-Kabana and Curl (1980), "...most pesticides act upon a great many organisms for which the chemicals were not specifically formulated, and this has resulted in altered biological phenomena which can be favorable or unfavorable to pathogen activity and disease development. We have seen that nematicides affect fungi, fungicides affect nematodes and nontarget fungi, and herbicides affect both of these as well as host physiology. A broad spectrum of chemicals is involved, including fumigant, nonfumigant, systemic, and nonsystemic compounds."

On the modern farm, a given hectare may be treated with a multitude of chemicals to control a variety of pests. The activity of these products can be altered further by individual formulation ingredients. Interactions that have been described include synergistic, additive, independent, and antagonistic (Putnam and Penner, 1974).

As farmers increasingly use combinations of pesticides, more research is needed on interactions among pesticides, on plant metabolism, and on soil chemistry and microflora (Simon-Sylvestre and Fournier, 1979).

OFF-SITE EFFECTS OF PESTICIDES

Conservation tillage increases water efficiency and reduces runoff and sedimentation, thereby improving water quality. To the degree that pesticides adhere to soil particles, transport to surface waters is reduced. If runoff produced by storms occurs while pesticide residues in the field are high, high concentrations are moved to surface waters, and several herbicides subsequently have been found in tap water (Baker, D., 1983). Effects of short-term high concentrations can be severe for fish and other aquatic organisms. For example, trifluralin in the unadsorbed state is relatively toxic to fish, and atrazine is toxic to immature stages of frogs (Mauck and Olson, 1976, unpublished).

Increased water soluble nutrient losses are experienced with conservation tillage due to incomplete incorporation of fertilizers with conservation tillage and to higher concentrations "in sediment from conservation tillage because the coarser, more dense, and least chemically active material is conserved to the greater degree" (Christensen and Norris, 1983). It has also been reported that the eroded particles from conservation tillage tend to consist of the more organically rich and valuable part of the soil. This may be "due to the selectivity of the erosion process in which finer, more chemically active sediment particles are preferentially transported" (Baker and Laflen, 1983).

Another means of transport of pesticides from target areas is frequently by drift or volatilization. If aerially applied, "the amount of pesticide transported to non-target areas by air may be as much as several hundred times as large as the amount transported by runoff" (Shoemaker and Harris, 1977).

Although surface pesticide loads to surface waters may be reduced, the overall load may be equal or higher. Even with pesticides applied in a granular formulation with ground equipment, such as carbofuran, runoff losses from no-till and conservation tillage corn grown on claypan soils were slightly higher than runoff losses observed with conventional tillage (Shoemaker and Harris, 1977). The same study pointed out that when greater quantities of pesticides are used, a greater amount of pesticide can be expected to be transported in runoff from no-tillage fields than from conventional tillage.

Recent monitoring studies have contradicted laboratory, field and modeling studies that concluded that leaching would not occur. Monitoring found increased percolation of water soluble pesticides that add to the load of chemicals finding their way to ground and surface waters. As many of the organochlorines have been removed from the market, farmers are left "with moderately adsorbed pesticides which are transported primarily in the dissolved phase rather than the solid phase of runoff" (Shoemaker and Harris, 1977). Non-adsorbed chemicals percolate downward in certain situations such as coarse-textured soil, cracks and crevices, and shallow aquifers.

Leaching potential of water-soluble agricultural chemicals under no-till is enhanced because of increased infiltration capacity of the soil, less evaporation and enhanced soil porosity. Leaching also occurs as channels through which water and nitrate move are undisturbed, promoting deeper soil penetration (Christensen and Norris, 1983).

Drawing attention to potential increases of agricultural chemicals in water is a 1984 study by the U.S. Geological Survey reporting that the trend toward minimal to no-till methods "is accompanied by a greater application of fertilizers and pesticides" which in time "may appear in nearby streams and may infiltrate into the shallow alluvial aquifers" [in Iowa].

In order to improve water quality and to prevent serious sedimentation and chemical pollution, greater attention needs to be paid to how chemicals move once applied to croplands, ways to improve technologies as now practiced, and ways to reduce quantities rather than increase them on our crops.

CROP ROTATIONS

Crop rotations diminished in the 1970s "because of development of new, efficient pesticides that permit continuous cropping of economically advantageous crops" (USDA-ARS, 1975). As crop rotations were abandoned, requests for more powerful pesticides were submitted to the Environmental Protection Agency. EDB for use on nematodes in soybeans in the Southeast was needed because rotation with grain sorghum was no longer economical. After groundwater contamination was detected and verified in Georgia, suspension of that use was ordered by EPA.

In 1975 a well-known scientist cautioned, "The basis for all effective systems must be rotation." The classical and time-honored sequences of recognized effectiveness has been rotation from row crops to pastures and back to row crops, or from crops to fallow. In Day's words (Day, 1978):

"The more drastic the rotation the greater the impact on the weed flora. For example, the rotation of rice lands to upland crops and back to rice is a powerful means of suppressing aquatic weeds on the one hand and terrestrial ones on the other. Changes in grazing or fallow procedures, in methods of irrigation or fertility practices, and other components of rotation schemes have their effects on weed floras."

Although herbicides initially occupied a supportive role to supplement tillage, with no-till, herbicides are totally dominant. With the development of precision machines and highly selective herbicides, crop rotation has declined even more. The farmer is thus locked into continuous cropping of economically advantageous crops.

It remains to be seen if farmers can use suitable rotation sequences to reduce insect and weed problems. Bill Hayes cautioned, "Crop rotation as a means of pest control is even more important when using minimum tillage systems than conventional tillage" [emphasis in original] (Hayes, 1982).

The environmental benefits of crop rotations are well known. As much as a 40 percent reduction on runoff losses or organochlorine insecticides was observed from a potato-oats-sod rotation compared with continuous potatoes.

In an analysis of energy use under different production methods and cropping systems, Doering and Peart found that pesticide costs could be reduced by 40 percent when crop rotations that include an alfalfa crop were used instead of a system where corn is grown continuously (Table 2).

In order to effectively utilize rotations for both environmental and economic benefits, several needs have been identified. Farmers should be able to rotate without penalty of reduced profits. The marketplace needs to encourage resting the land or planting a lower-priced crop. There is no current incentive for chemical companies to recommend rotation of chemicals. The impetus is obviously

Table 2. Pesticide Costs per Hectare (Doering and Peart, 1977)

Crop	Total Cost	Herbicides	Insecticides
Continuous corn	$44.46	$27.17	$17.29
Corn-soybeans-wheat- alfalfa	26.18	11.86	14.33

to keep the immediate profit line up for each chemical rather than a group of chemicals, or the long-term market of a single chemical. Machinery could be made more flexible to accommodate crop and chemical rotations.

CONCLUSION AND OUTLOOK

Conservation tillage should not be looked upon "as a permanent and complete shift to an entirely new system of crop production," wrote Doug Worsham. It "should be looked upon as a management practice that can be incorporated where most advantageous in a grower's program" (Worsham, 1980).

We have time to do this right. In W.E. Larson's words, "Erosion is a serious problem that needs to be dealt with. But we are not facing catastrophic losses of all production in the next few years. We have the time to make a reasoned study of all inputs and costs and then take the needed steps in an orderly fashion." USDA now has data on where erosion is threatening productivity and where cropping should not be continued. Larson's preliminary analysis of USDA data leads him to predict that losses in productivity would be the greatest on slopes exceeding 6 percent and when subsoil properties are unfavorable for crop growth. He also estimates that about 20 percent of our cropland is on slopes steeper than a 6 percent grade (Larson, 1983).

In the words of another participant at this conference, "Our challenge is to develop tillage systems and other erosion-control methods for these problem soils where an erosion hazard exists. We need to be more imaginative and innovative in offering the farmer a wider range of alternative methods for controlling erosion" (Moldenhauer, 1979).

In bringing together such a broad spectrum of experts, and other interested parties, this symposium initiates an important process of identifying research needs for conservation tillage management strategies. As a systems approach is put together, let me conclude with a statement by Carl Huffaker: "delay is not necessary while extensive studies are undertaken. Basic individualistic and practical studies are...a part of [and] must both precede and run concurrently with other approaches utilizing an analysis of the whole system, with stress on main areas and promising tactics...Systems analysis is used...to point to gaps in research findings, the two approaches...go hand in hand...The problems posed when we consider all aspects of crop production...are too complex to handle reliably without the aid of a quantitative, patterned procedure for decision-making. The farmer has had to be his own systems analyst. Scientists must do a better job for him. With luck, good coordinated ecological research from all relevant disciplines, and an eye always to the practical, we can do so" (Huffaker, 1974).

SUMMARY

This paper discusses a variety of environmental concerns regarding pesticides and their impacts on croplands, comparing tillage practices. The possibility of increased pesticides involved in conservation tillage is discussed. Several research needs are identified as an outgrowth of such increases in quantities and different types of chemicals. Also explored is the assumption that chemical pollution of surface and groundwaters may be reduced with conservation tillage. There is some evidence that chemical pollutants may increase with reduced tillage systems. The author emphasizes crop rotations as a means to reduce both chemical pollution and reduce reliance on increased chemicals to meet certain pest problems. There is time to proceed intelligently to undertake studies required by the complexity of crop production, and to develop a systems approach to address management problems involved in conservation tillage. More emphasis on the entire management system will make possible rotations, cover cropping, including living mulch, and multiple cropping. The end result should help the farmer.

LITERATURE CITED

Baker, D.B. Herbicide contamination in municipal water supplies of northwestern Ohio. Draft Final Report. October 2, 1983. 33 pp.

Baker, J.L. and J.M. Laflen. 1983. Water quality consequences of conservation tillage. J. Soil Water Cons. 38:187-193.

Christensen, L.A. and P.E. Norris. 1983. A comparison of tillage systems for reducing soil erosion and water pollution. U.S. Dept. of Agriculture Economic Research Service. Agricultural Economic Report Number 499. May. 27 pp.

Crosson, P. 1981. Conservation tillage and conventional tillage: A comparative assessment. Soil. Cons. Soc. Am., Ankeny, Iowa. 35 pp.

Day, B.E. 1978. The status and future of chemical weed control. In: Pest Control Strategies. E.H. Smith and D. Pimental (eds.). Academic Press, New York, NY. pp. 203-213.

Doering, O., III and R.M. Peart. 1977. Evaluating alternative energy technologies in agriculture. RANN Rpt. No. 770124. Nat. Sci. Found., Washington, D.C. 13 pp.

Epplin, F.M., T.F. Tice, A.E. Baquet, and S.J. Handke. 1982. Impacts of reduced tillage on operating inputs and machinery requirements. Am. J. of Agr. Economics, Vol. 64, #5. pp. 1039-1046.

Hayes, W.A. 1982. Minimum tillage farming. No-Till Farmer, Inc., Brookfield, WI. 166 pp.

House, G.J. and B.R. Stinner. 1983. Arthropods in no-tillage soybean agroecosystems: Community composition and ecosystem interactions. Environmental Management. 7(1), pp. 23-28.

Huffaker, C.B. 1974. Some ecological roots of pest control. Entomophaga. 19(4):371-389.

Kaufman, D.D. and P.C. Kearney. 1976. Microbial transformations in the soil. In: Herbicides. L.J. Audus (ed.). Academic Press, New York, NY. pp. 29-64.

Larson, W.E., F.J. Pierce and R.H. Dowdy. 1983. Soil erosion and productivity, a rational look at a complex situation. Crops and Soils. Aug.-Sep. pp. 19-20.

McWhorter, C.G. 1978. Disciplinary challenges: Weed science. In: Development of Optimum Crop Production Systems for the Mid-South. Ark. Agr. Exp. Sta., Fayetteville. pp. 41-44.

Moldenhauer, W.C. 1979. Erosion control obtainable under conservation practices. In: Universal Soil Loss Equation: Past, Present, and Future. Soil Science Society of America Special Publication Number 8, SSSA, Madison, WI. pp. 33-43.

Moore, E.G. 1967. The agriculture research service. Frederick A. Praeger, New York, NY. pp. 188-222.

Putnam, A.R. and D. Penner. 1974. Pesticide interactions in higher plants. Residue Rev. 50:73-110.

Rodriquez-Kabana, R. and E.A. Curl. 1980. Nontarget effects of pesticides on soilborne pathogens and disease. Phytopathol. 18:311-32.

Shoemaker, C.A. and M.O. Harris. 1979. The effectiveness of SWCPs in comparison with other methods for reducing pesticide pollution. In: Effectiveness of Soil and Water Conservation Practices for Pollution Control, Douglas A. Haith and Raymond C. Loehr (eds.). Environ. Prot. Agency. Athens, GA. pp. 206-324.

Simon-Sylvestre, G. and J.C. Fournier. 1979. Effects of pesticides on the soil microflora. Advances in Agron. 31:1-92.

United States Department of Agriculture, Agriculture Research Service and Environmental Protection Agency Office of Research and Development. (USDA-ARS and EPA-ORD). 1975. Control of water pollution from cropland. Vol. 1, Rpt. No. EPA-60012-75-026 or Rpt. No. ARS-#-5-1. 111 pp.

United States Geological Survey. 1984. National water summary 1983--hydrologic events and issues. Water-Supply Paper 2250. 243 pp.

Worsham, A.D. 1980. No-till corn--its outlook for the 80s. 35th Annual Corn & Sorghum Research Conference. pp. 146-163.

CHAPTER 23

CONSERVATION VS. CONVENTIONAL TILLAGE:
WILDLIFE MANAGEMENT CONSIDERATIONS

Louis B. Best
Department of Animal Ecology
Iowa State University
Ames, Iowa 50011

INTRODUCTION

Unlike some disciplines in agriculture such as agronomy, plant science, and economics, where research on conservation tillage has been underway for over 40 years (IAHEES, 1981), attempts to quantify effects of various tillage practices on wildlife have been initiated only within the last few years. In part, this has resulted because wildlife biologists and managers traditionally have considered land used for row crop production to be of relatively little value for wildlife, particularly when compared with areas covered by native vegetation, forage crops, or small grains. To may wildlifers, the Corn Belt is instead the "corn desert." We know very little about the ways in which wildlife depend upon and use conventionally tilled cropland, and even less about the effects of various conservation tillage practices. Our knowledge for game species, such as the ring-necked pheasant (Phasianus colchicus) is scant, and for nongame wildlife, it is essentially nonexistent. In some regions, the proportion of land devoted to row crops is extremely large, and despite the seemingly low value of this land for wildlife, some species depend on it for their very survival. In some cases it is the only alternative available.

The terms "conservation tillage" and "conventional tillage" connote different things to different people. Therefore, I will give my working definitions of the terms. Conservation tillage shall be defined as a tillage system that does not involve inverting the soil with a moldboard plow, and one in which crop residue generally remains on the soil surface. The amount of residue left on the soil surface depends upon the particular tillage operation used and the previous crop grown. Conventional tillage refers to a system where the surface soil and crop residue are turned under with a moldboard plow, usually in the fall, followed by disking and harrowing the next spring, thus creating a smooth, residue-free surface.

Although forms of conservation tillage are practiced throughout the Great Plains and Great Lakes regions, I will restrict my discussion to those areas where the principal crops grown are corn and soybeans, and where continuous row cropping is the prevailing farm operation. Thus, my remarks will be most relevant to states such as Iowa, Illinois, and Indiana. The effects of conservation tillage in more western regions (e.g., Cowan, 1982; Rogers, 1983) where small-grain production predominates will not be addressed. References to wildlife will be confined to species that are primarily terrestrial in habit; effects of tillage practices on aquatic environments will not be discussed.

A quick perusal of most textbooks on wildlife conservation or management (e.g., Leopold, 1933; Giles, 1978) would reveal that certain requisites must be present in order for wildlife to thrive and sustain population numbers. Environmental requirements most frequently discussed are food, cover, water, and interspersion. With food, important considerations include food quantity, quality, and availability. Corn is considered one of the leading wildlife foods in the United States (Martin et al., 1951), and it is an important energy source for many wildlife species (e.g., ring-necked pheasants and waterfowl along their migration corridors) during fall, winter and spring. Cover can serve several functions, including shelter from inclement weather (particularly during the winter), protection or escape from predators, and nesting or resting sites. Water is an important factor limiting wildlife populations in some arid regions, but it is of minor importance in our region and, therefore, will not be discussed further. Interspersion refers to the juxtaposition of different habitat types. For example, an abundant food supply may be of little value to wildlife if needed cover is not nearby. In an agriculture setting, interspersion is influenced by field size and the diversity of crops grown. I will focus primarily on the food and cover needs of wildlife, with briefer consideration of interspersion.

My approach will be to discuss the extent to which wildlife requisites are satisfied in conventional tillage systems and then to evaluate whether or not conditions are improved through conservation tillage. I will consider first winter habitat requirements and then summer habitat requirements. Contrasts between conservation and conventional tillage that affect wildlife most are differences in (1) the amount of residue cover, (2) the availability of waste grain on the soil surface, (3) the frequency of passes over the field with farm machinery, and (4) the means used to control weeds. After considering wildlife habitat requirements, I will discuss potential wildlife depredation problems associated with conservation tillage.

RESULTS AND DISCUSSION

Winter Habitat Requirements

During late fall, winter, and early spring, food and/or cover often limit wildlife populations on conventionally tilled cropland. Dependency on cropland food and cover differes among wildlife species; those using cropland to the greatest extent are most impacted by tillage practices. Fall tillage with the moldboard plow can bury as much as 95 percent of the corn residue beneath the soil surface, and with it, most (99 percent) of the waste grain (Warner et al., 1985). Other fall tillage practices, such as chisel or disk plowing, are less detrimental to wildlife than using the moldboard plow, but they are notably inferior in terms of winter habitat to systems such as slot planting or till planting, where fall tillage is avoided altogether. As a general rule, any tillage practice that leaves more crop residue on the soil surface, particularly standing residue, will enhance the cover value of cropland during the winter for wildlife. The relative importance of crop residue in providing wildlife cover is influenced by the severity of the winter weather and the proximity of alternate cover (e.g., woodlots, fencerows, windbreaks) to the food source--the cropland. The interspersion of food and cover often determines whether or not wildlife can utilize a food supply in crop fields.

Many factors may influence the availability of waste grain on cropland. The proportion of waste grain buried by different tillage practices varies greatly (Warner et al., 1985), and hence, so does availability to wildlife. In general, tillage practices that leave more residue on the soil surface also leave more waste grain. Tillage operations also influence the form of the waste grain--that is, whether or not the corn is on the cob. Baldassarre et al. (1983), reported that 70 percent of the corn was ear waste and 30 percent kernel waste in freshly harvested fields. This may be unimportant to larger game birds, such as geese and ring-necked pheasants, but smaller birds [e.g., teal (Anas spp.) and mourning doves (Zenaida macroura)] may be unable to remove kernels from the cob. More total waste corn remains on the soil surface with no fall tillage than after disking, but disking shatters corn ears into smaller pieces, thus increasing availability of the remaining waste corn (Baldassarre et al., 1983). Often livestock are grazed on cropland after harvest, and grain supplies may be reduced substantially (e.g., 84 percent reduction; Baldassarre et al., 1983) depending on the number of animal units per hectare and how long they remain on the field. Farmers may graze livestock on fields where waste is most abundant (Baldassarre et al., 1983), thus reducing the difference in waste grain availability between fall-plowed fields and those remaining untilled. But roaming cattle knock kernels off cobs similar to disking, thereby increasing availability of the unconsumed corn. Snow or ice cover also may render waste grain unavailable to wildlife. With more intensive grazing of crop fields and/or with greater snow/ice cover, wildlife become more dependent on conservation tillage practices that assure an adequate supply of food on the soil surface. But no-tillage fields may trap and hold blowing snow better than conventionally tilled fields, thus making a more abundant food source less available. During particularly severe winters with deep snow, conservation tillage may offer no advantage to wildlife over conventional tillage.

Despite the importance of food and cover for wildlife during the winter, very few studies have quantified the effects of different tillage systems on wildlife abundance or food availability. Recently, Castrale (1983a) reported results from a study in Indiana that evaluated winter use of corn and soybean fields by birds and small mammals. Fall tillage treatments in his study included disking, chisel plowing, moldboard plowing, and no tillage. Two small-mammal species were found commonly in the crop fields. Deer mice (Peromyscus maniculatus) were most abundant in fields with reduced crop residue, and house mice (Mus musculus) were most consistently found in corn fields with high crop residue. Castrale's data suggest that house mice are negatively impacted by tillage and must recolonize fields after disturbances. Deer mice, however, can persist and even flourish under intense cultivation.

Thirty-three bird species were observed using the study fields during the winter (Castrale, 1983a). Species most frequently encountered included horned lark (Eremophila alpestris), eastern bluebird (Sialia sialis), mourning

dove, American crow (Corvus brachyrhynchos), killdeer
(Charadrius vociferus), dark-eyed junco (Junco hyemalis),
and song sparrow (Melospiza melodius) (Castrale and Speer,
1984). Untilled corn and soybean fields supported the
highest densities and greatest variety of birds. Fields
tilled with moldboard or chisel plows were rarely used
by birds, and disked corn fields received intermediate
bird use. The only two bird species using plowed fields
(horned lark and killdeer) are known to prefer open habitats
with reduced cover. Searching for food was the primary
use of crop fields by most birds, although mourning doves
roosted in field residues at night. In many cases, use
of crop fields was influenced by the fields' proximity
to preferred habitats containing more woody or herbaceous
cover. Fields, even those with high residue amounts,
seldom were used by most birds for escape cover; primary
exceptions included killdeer, mourning dove, horned lark,
and eastern bluebird.

In Iowa (Basore, 1983), very few bird species were
observed during brief censuses of corn and soybean fields
in winter. In contrast to the 33 species recorded in
crop fields during the winter in Indiana (Castrale, 1983a),
only four were observed in Iowa. Differences in the amount
of snow cover, number of winter resident bird species,
and intensity of the censuses probably account for most
of the disparity in the results from the two studies.

In addition to direct effects on food availability
and cover in crop fields, fall cultivation also can influence
the suitability of strip cover adjacent to cropland.
When fields are tilled in the fall, soil and snow from
the cropland may be blown into adjacent roadside ditches,
thus burying important strip cover that would otherwise
be available to wildlife. Soil deposition is most extreme
during winters with limited snow cover and when soil moisture
is low.

Not only does the type of primary tillage affect
availability of food and cover for wildlife in cropland,
but also the season when the tillage occurs. Spring tillage
is vastly superior to fall tillage because, although food
and cover may be reduced during the spring and summer
breeding season, these wildlife resources are maintained
at higher levels during the winter season when the need
is most critical.

Summer Habitat Requirements

The environmental factors limiting wildlife populations
in the late spring, summer, and early fall differ substantially

from those most important during the winter. Food becomes more abundant as insects emerge and their populations increase in size. Most species of birds breeding in cropland consume a higher proportion of insects in their diets during summer than during winter (Martin et al., 1951). Waste grain is less available (due to decomposition, tillage, consumption by wildlife, etc.), but weed seeds become more abundant as the season progresses. Consequently, food availability generally is less limiting to wildlife in summer than in winter.

Fall tillage practices that influence availability of waste grain during winter or primary tillage in spring continue to influence the amount of grain remaining on the soil surface through the end of summer. In addition, ground preparation for planting in spring and subsequent cultivation have an impact; practices such as slot-planting, that result in minimum tillage, preserve the greatest quantity of grain for wildlife. Corn kernels are more abundant during summer than soybeans because the latter decompose more rapidly (Warner et al., 1985). Some researchers have reported a positive relationship between total insect abundance and crop residue (IAHEES, 1981; Blumberg and Crossley, 1983), but Basore (1984) found no such relationship during the brood-rearing period of pheasants. Wildlife usage of insect foods, however, depends not only on the quantity but also the quality; certain insect taxa are preferred over others. Conservation tillage probably changes the species composition and/or relative abundance of individual species within the insect communities (e.g., Blumberg and Crossley, 1983). Unfortunately, the effects of various tillage practices on wildlife food habits have not been vigorously studied. Young (1984) has documented small-mammal diets in no-tillage and disked corn fields.

Cover remains important to many wildlife species during summer, but it serves different functions than in winter. Late spring and summer constitute the breeding season for many species of wildlife, particularly birds, and vegetative cover provides concealment for nest sites and protection against inclement weather. Cover also continues to be important as avenues of escape from predators.

The tillage system employed to prepare the ground for planting, and subsequent measures used to control weeds, impact on birds breeding in crop fields in several important ways. They influence the amount of crop residue available for nesting cover, the frequency of passes over the field with farm machinery, and the types of chemicals to which wildlife are exposed.

Recent studies in Iowa (Basore, 1984) and Indiana (Castrale, 1983b) have shown that no-tillage crop fields have higher densities and a greater variety of birds during the breeding season than conventionally tilled fields. In Iowa, species found nesting in no-tillage fields included vesper sparrow (Pooecetes gramineus), killdeer, mourning dove, ring-necked pheasant, grasshopper sparrow (Ammodramus savannarum), western meadowlark (Sturnella neglecta), dickcissel (Spiza americana), and field sparrow (Spizella pusilla). In addition to those bird species nesting in cropland, many other species visit crop fields to feed, rest, etc.; this is particularly true during spring and fall migration. It seems clear that the major factor attracting more birds to no-tillage fields is the greater amount of crop residue on the soil surface.

One benefit of conservation tillage frequently extolled is the reduction in fuel costs associated with making fewer passes over the crop fields. Reducing field operations also may benefit wildlife because each time farm machinery is used in a field, nests may be destroyed and the general vicinity of other nests may be disturbed such that the nests are abandoned. The severity of this on nesting birds depends upon where the nest is positioned relative to the crop row, the duration of the nesting cycle, the tendency to renest after nesting failure, and the timing of the breeding season. Some bird species show a propensity to place the nest either within or between crop rows, the location probably depending upon the distribution of crop residue and the stage of development of crop plants. Nests between crop rows have the greatest likelihood of being destroyed either by the tractor tires or the farm implement. Birds' nesting cycles also differ in length; nesting cycles (period from egg laying to the young leaving the nest) of the species known to nest in corn and soybean fields (Basore, 1984) vary from 21 to over 35 days (calculated from data in Harrison, 1978). The shorter the nesting cycle, the greater the likelihood of successfully raising young between successive passes over a field with farm machinery. Conservation tillage decreases the frequency of field operations and probably provides the greatest benefit to those species with long nesting cycles. The propensity to renest after nest failure also influences the effects of tillage practices on the reproductive output of farmland birds. Again, conservation tillage would be most beneficial to species that raise only single broods or that renest infrequently. The ring-necked pheasant is a single-brooded species (although some will renest), and it also has the longest nesting cycle of the species that nest in crop fields. This makes it particularly susceptible to frequent implement passes over the field.

Some bird species (horned lark) initiate nesting before field operations begin in the spring; the breeding seasons of others (e.g., mourning dove and vesper sparrow) extend beyond the period when crops normally are cultivated. Species whose nesting season is confined to the period when most field operations occur stand to benefit most from conservation tillage practices that require fewer passes over the field.

A study in Iowa (Young, 1984) reported greater small-mammal species diversity in no-tillage fields compared to disked fields, although the total small-mammal abundances did not differ. Deer mice dominated small-mammal communities on all fields; thirteen-lined ground squirrels (Spermophilus tridecemlineatus) were second in importance. Results from this study suggest that increased residue cover tends to diversify rather than increase abundance of small mamals.

In recent years, there has been a general trend toward greater reliance on chemicals for weed control. By its very nature, conservation tillage (particularly slot planting) has necessitated a shift from mechanical cultivation to chemical weed control. In addition to its influence on the number of passes over a field with farm machinery and the degree of ground disturbance associated with each pass, use of herbicides may present a more direct threat to wildlife because of toxic effects. The greatest risk is to developing embryos in eggs and to nestling birds, which, because of their immobility, may be unable to avoid contact with herbicide spray. Herbicides differ in their toxicity to wildlife; some seemingly are harmless (except for indirect effects on the habitat) while others have proven to be highly toxic (Dunachie and Fletcher, 1970; Batt et al., 1980; Hoffman and Eastin, 1982). Paraquat, a contact "burndown" herbicide used before planting to kill early weed or sod growth, is particularly harmful to wildlife. The method of applying herbicides also may influence the herbicides' effects on wildlife. Incorporated herbicides pose less threat to wildlife than those surface applied, and herbicides applied outside the breeding season are less damaging. Some forms of conservation tillage, such as slot-planting, require surface application. Much more research is needed before we will fully appreciate the impacts of herbicides on wildlife.

There is abundant evidence that insecticides adversely affect wildlife populations. But because insecticide use relates more to cropping sequences than to tillage practices (Stockdale, 1983), it will not be discussed further.

Potential Crop Depredation by Wildlife

A concern frequently voiced about conservation tillage is that it invites greater rodent damage to developing corn seedlings. In Iowa, reported crop damage has been sporadic throughout the state as well as within individual crop fields; the most severe damage by rodents has occurred on experimental study plots (Moorman, 1983). By design, these plots are small and usually bordered by mowed grass strips, thus creating an idea edge habitat for thirteen-lined ground squirrels--the major cause of rodent damage to corn seedlings (Johnson et al., 1982). A study of field-sized plots in Iowa (Young, 1984) recorded only minor crop damage by rodents. (Damage by birds was negligible.) Corn seedling mortality due to rodents was 0.16 percent in disked fields and 0.12 percent in no-tillage fields planted in corn residue. Such low levels of crop damage by rodents, if substantiated by other studies, could easily be offset by increasing seeding rates slightly, thus avoiding the use of rodenticides or repellants. In addition, rodents feed heavily on lepidopterous larvae and weed seeds during summer and fall (Houtcooper, 1977), a factor that could compensate, at least in part, for their damage to crop plants. In instances where rodent damage within large fields is extensive, the adequacy of the weed control program should be evaluated before initiating a rodent control program. Evidence suggests that ground squirrels are attracted more by grass cover than by crop residue. In the Iowa study, corn seedling mortality in fields with sod residue was over 100 percent greater than that in fields with corn residue (Young, 1984). This also illustrates the influence of cropping sequence on the potential for rodent-induced crop damage. Crop damage in conservation tillage systems can be expected to be higher when pasture land or hay fields are first seeded into corn than in a continuous row-cropping operation.

SUMMARY

Only recently have the effects of various tillage practices on wildlife been quantified. Contrasts between conservation and conventional tillage that most affect wildlife are differences in the amount of crop residue cover, availability of waste grain, frequency of passes over the field with farm machinery, and means used to control weeds. During the winter, food and cover often limit wildlife populations on cropland; conservation tillage increases the availability of these habitat requisites. Few small-mammal species over-winter in corn and soybean fields. Deer mice thrive in fields with little crop residue; house mice require more ground cover. Untilled corn and soybean fields support the highest densities and greatest variety of birds during the winter; plowed fields are essentially unused. During summer, availability of suitable nesting cover limits breeding populations of birds on cropland. No-tillage crop fields support higher densities and a greater variety of nesting birds than do conventionally tilled fields. Vesper sparrow, killdeer, mourning dove and ring-necked pheasant are the most common species nesting in no-tillage fields. Reduced tillage lessens the risk of destroying nests with farm machinery. Increased use of herbicides associated with conservation tillage may adversely affect wildlife depending on mode of application and relative toxicity. Rodent damage to developing seedlings seems influenced more by field size, cropping sequence, and completeness of the weed control program than by the tillage system used.

ACKNOWLEDGMENTS

A previous draft of this paper was reviewed by N. Basore, R. Cruse, J. Dinsmore, and R. Young; their helpful comments are appreciated.

LITERATURE CITED

Baldassarre, G.A., R.J. Whyte, E.E. Quinlan, and E.G. Bolen. 1983. Dynamics and quality of waste corn available to postbreeding waterfowl in Texas. Wildl. Soc. Bull. 11:25-31.

Basore, N.S. 1983. Unpublished data. Dept. of Animal Ecology, Iowa State Univ., Ames, IA.

Basore, N.S. 1984. Breeding ecology of upland birds in no-tillage and tilled cropland. M.S. Thesis, Iowa State Univ., Ames, IA. 62 pp.

Batt, B.D.J., J.A. Black, and W.F. Cowan. 1980. The effects of glyphosate herbicide on chicken egg hatchability. Can. J. Zool. 58:1940-1942.

Blumberg, A.Y. and D.A. Crossley, Jr. 1983. Comparison of soil surface arthropod populations in conventional tillage, no-tillage and old field systems. Agro-Ecosystems 8:247-253.

Castrale, J.S. 1983a. Fall agricultural tillage practices and field use by wintering wildlife in southeastern Indiana. Pages 53-54 in Abstr. 45th Midwest Fish Wildl. Conf., St. Louis. 142 pp.

Castrale, J.S. 1983b. Wildlife use of conservation tillage study fields, summer 1983. Wildl. Manage. Res. Notes, No. 230, Indiana Dept. Nat. Resour. 2 pp.

Castrale, J.S., and R.T. Speer. 1984. Wintering birds of corn and soybean fields in southeastern Indiana. Amer. Birds 38:57-59.

Cowan, W.F. 1982. Waterfowl production on zero tillage farms. Wildl. Soc. Bull. 10:305-308.

Dunachie, J.F., and W.W. Fletcher. 1970. The toxicity of certain herbicides to hens' eggs assessed by the egg-injection technique. Ann. Appl. Biol. 66:515-520.

Giles, R.H., Jr. 1978. Wildlife Management. W.H. Freeman and Co., San Francisco. 416 pp.

Harrison, C. 1978. A Field Guide to the Nests, Eggs and Nestlings of North American Birds. Collins, New York. 416 pp.

Hoffman, D.J., and W.C. Eastin, Jr. 1982. Effects of lindane, paraquat, toxaphene, and 2,4,5-trichlorophenoxyacetic acid on mallard embryo development. Arch. Environ. Contam. Toxicol. 11:79-86.

Houtcooper, W.C. 1977. Food habits of rodents in a cultivated ecosystem. J. Mammal. 59:427-430.

IAHEES. 1981. Our thinning soil: an update on conservation tillage. Research for a Better Iowa, AR-14. Iowa Agriculture and Home Economics Experiment Station, Iowa State Univ., Ames, IA. 5 pp.

Johnson, R.J., A.E. Koehler, and O.C. Burnside. 1982. Rodent repellents for planted grain. Pages 205-209 in R.E. Marsh, ed. Proc. 10th Vert. Pest Conf., Univ. California, Davis.

Leopold, A. 1933. Game Management. Charles Scribner's Sons, New York. 481 pp.

Martin, A.C., H.S. Zim, and A.L. Nelson. 1951. American Wildlife and Plants. McGraw-Hill Book Co., Inc., New York. 500 pp.

Moorman, R. 1983. Personal communication. Dept. of Animal Ecology, Iowa State Univ., Ames, IA.

Rodgers, R.D. 1983. Reducing wildlife losses to tillage in fallow wheat fields. Wildl. Soc. Bull. 11:31-38.

Stockdale, E. 1983. Personal communication. Dept. of Entomology, Iowa State Univ., Ames, IA.

Warner, R.E., S.P. Havera, and L.M. David. 1985. Effects of autumn tillage systems on corn and soybean harvest residues in Illinois. J. Wildl. Manage. 49:185-190.

Young, R.E. 1984. Response of small mammals to no-till agriculture in southwestern Iowa. M.S. Thesis, Iowa State Univ., Ames, IA. 66 pp.

CHAPTER 24

FARMERS' ATTITUDES AND BEHAVIORS IN IMPLEMENTING CONSERVATION TILLAGE DECISIONS

Peter J. Nowak
University of Wisconsin - Madison
Madison, Wisconsin 53706

INTRODUCTION

The rapid diffusion of conservation tillage has been called the latest revolution in American agriculture (Hinkle, 1983). Depending on the estimate cited, there has been a shift from between 3.2 (Block, 1983) and 17.8 (Lessiter, 1974) million ha in some form of conservation tillage in 1973 to between 38 (Conservation Tillage Information Center, 1983) and 51 (Lessiter, 1983) million ha in 1983. Predictions on the future of this trend are also optimistic. In the year 2000 it is expected that conservation tillage will be used on approximately 60 percent of our cropland according to Crosson (1981), near 70 percent according to the Office of Technology Assessment (1982), and around 80 percent according to USDA (1975) estimates. Others even foresee complete adoption (Kinney, 1983). From only a few farmers using this technology several years ago, there are claims that between 75 (Pioneer Hy-bred International, Inc., 1982) and 80 (Crosson, 1982) percent of all farmers employ it today. Based on all these encouraging reports one could conclude that conservation tillage has indeed revolutionized our approach to soil and water conservation.

However, before we unsaddle the conservation warhorses and pat ourselves on the back for promoting this latest technological fix to an agricultural problem, we need a closer examination of these "revolutionary" facts. The premise of this paper is that the diffusion of conservation tillage among farm audiences has been overestimated in scope due to underestimating the magnitude and complexity of change required. In particular, it is argued that the diffusion of conservation tillage is an evolutionary and not revolutionary process.

The Diffusion of Conservation Tillage

The diffusion of conservation tillage has and will continue to be influenced by three conditions. The first condition is the development of supporting technologies. For example, diffusion rates within specific ecological settings have been highly dependent on the development of pesticides to counter new weed or pest conditions. Agricultural machinery has had to be either modified or created in response to more challenging operating conditions. There has also been the development of new seed varieties in response to a different growing environment. The diffusion of conservation tillage in different regions of the country has been influenced by the rate of innovation of these supporting technologies.

The second condition relates to how these supporting technologies are distributed among the farm population. The generation and distribution of knowledge are discrete processes. Although the support of private industry has been uniform for the profitable supporting technologies, there is tremendous variation in how government and conservation organizations promote the adoption of conservation tillage. Some areas have made strong efforts to make these supporting technologies available to all farmers in a practical and timely manner. Other areas still believe it is the farmer's responsibility to either generate or locate answers to tillage problems.

The third condition relates to the social and economic factors influencing the individual's adoption decision. Farmers' decisions relative to conservation tillage are strongly influenced by their social communities, institutional support, and market conditions (Nowak and Korsching, 1983). The adoption decision is neither trivial nor simple. The farmer's ability to overcome the obstacles to adopting a conservation tillage system is largely dependent on these socioeconomic factors (Nowak, 1983b). The rate of diffusion within agricultural communities is also related to these factors. Successful promotion of conservation tillage has been based on the understanding of this process.

Inability to completely control these conditions has prevented a conservation tillage revolution from occurring in American agriculture. Instead the diffusion of this technology continues in an evolutionary manner.

THE STUDY

The study was part of an interdisciplinary effort exploring the effect of agricultural land-use practices on stream water quality. The sociological component involved three watersheds in the Iowa-Cedar River Basin of east-central Iowa. They were selected on the basis of matched soil, topographic, and socioeconomic characteristics of the farm firms. All persons owning or operating land in these watersheds were interviewed. Data on 193 farm operators were collected during 1980 and 1981. Factors related to personal and socioeconomic characteristics of the farm operator, farm enterprise features, ecological characteristics, agronomic patterns, and institutional involvement were measured. From this information, crop residue was calculated on a field by field basis (Nowak and Korsching, 1984).

RESULTS

Reported Versus Actual Use of Conservation Tillage

If the farmers have adopted the tools but not the techniques of conservation tillage, then one can expect a number of farmers who will report adoption even though they are not maintaining adequate residue levels necessary for conservation objectives (Table 1). Respondents were asked if they were using conservation tillage in their operation. Of the 193 respondents, 151 (78.2 percent) claimed to be using a conservation tillage system. Actual residue levels were then estimated from tillage-induced reductions on a field by field basis. There were 135 respondents who provided complete agronomic information on their fields and who had some fields coming out of a corn rotation. Assuming a liberal interpretation of conservation tillage to be an average of 1680 kg of corn residue per ha prior to spring planting, then only 26 or 19.3 percent of the 135 were actually practicing conservation tillage on these fields.

There were 110 respondents who provided agronomic information for fields coming out of a soybean rotation. Assuming a liberal interpretation of conservation tillage to be an average of 840 kg of soybean residue per ha prior to spring planting, then 52 or 47.3 percent of the 110 were actually practicing conservation tillage on these fields.

When using the more conservative standards of 2240 kg/ha of residue for corn and 1120 kg/ha of residue for soybeans, then only 7.4 percent (N=10) of the farmers

Table 1.   Reported Versus Actual Use of Conservation Tillage

---

Reported Use of Conservation Tillage

Using Conservation Tillage	78.2% (151)[1]
Not Using Conservation Tillage	21.8% (42)

Actual Residue Levels[2] For Those Reporting Use of Conservation Tillage

Corn Residue		Soybean Residue	
1680 kg/ha	19.3% (26)	840 kg/ha	47.3% (52)
2240 kg/ha	7.4% (10)	1120 kg/ha	26.4% (28)

---

1. Number of respondents is indicated within ( ).
2. Average for all fields with same rotation. Estimated residue levels based on USLE calculations and prior to spring planting.

were actually practicing conservation tillage on their corn fields, and 26.4 percent (N=29) were actually practicing conservation tillage on their soybean fields.

Obstacles to Conservation Tillage

Quite clearly there is a significant difference between reported and actual use of conservation tillage among farmers. Various obstacles prevent the operator from using a conservation tillage system to its fullest capabilities. Two frequently mentioned obstacles are associated with weed control and fertilizer management (Conservation Tillage Information Center, 1983).

Respondents were asked the extent of the problem they had in matching herbicide and fertilizer management programs to conservation tillage systems. Problem perceptions were then associated with their actual residue levels. In Table 2 there is not the expected finding that those with the least problems also maintain the highest residue levels. Instead, in three of the four situations those with higher residue levels also believe that herbicide or fertilizer management is a moderate or serious problem. This indicates that although they were able to maintain higher residue levels, they also recognized a number of pesticide and fertilizer problems in doing so. Also contributing to these findings are the carryover problems in a corn–soybean rotation which would not be present in continuous corn.

Table 2.   Problems   in   Implementing   Conservation   Tillage
by Residue Levels

	Corn Residue (Kg/ha)	Soybean Residue (kg/ha)
Knowing How to Match Herbicides		
Not or only a Slight Problem	1040 (92)[1]	865 (78)
Moderate or a Serious Problem	710 (28)	1037 (22)
Knowing How to Match Fertilizer		
Not or only a Slight Problem	953 (102)	827 (85)
Moderate or a Serious Problem	1055 (18)	1294 (15)

[1] Number of respondents is indicated within ( ).

The   availability   of   effective   post-emergence   herbicides
for corn is another important consideration.

In   Table   3   perceptions   of   conservation   tillage   were
compared to actual residue levels.   Relative to the perceived
profitability   of   conservation   tillage,   the   majority   feel
that returns exceed costs.   However, also note that perceptions
of   profitability   are   weakly   associated   with   actual   residue
levels.   This   is   especially   true   relative   to   soybean   residue.
Profitable   conservation   tillage   systems   are   not   being
identified with higher residue levels.

The   statistics   associated   with   the   perception   of
time   and   labor   challenges   the   often   repeated   statement
that   the   adoption   of   conservation   tillage   will   mean   less
time   and   labor.   For   the   majority   it   does,   although   not
at   the   highest   residue   levels.   Note   that   the   highest
residue   levels   are   being   maintained   by   those   who   believe
that   conservation   tillage   involves   more   time   and   labor.
The   perception   of   no   change   in   time   and   labor   is   probably
accurate   in   that   the   residue   levels   indicate   a   change
in tillage tools, but not tillage practices.

The   majority   of   operators   felt   that   conservation
tillage   was   compatible   to   their   operation.   Those   perceiving
the   lowest   compatibility   also   maintained   lower   levels
of crop residue.

Table 3. Perceptions of Conservation Tillage by Residue Levels

Characteristics of Conservation Tillage	Corn Residue (kg/ha)	Soybean Residue (kg/ha)
Profitability		
Costs Exceed Returns	880 (11)[1]	936 (10)
Costs Equal Returns	750 (22)	883 (20)
Returns Exceed Costs	1204 (78)	983 (60)
Time and Labor		
More Time and Labor	1332 (12)	1025 (11)
No Change	538 (9)	861 (8)
Less Time and Labor	1103 (90)	955 (71)
Compatibility		
Low Compatibility	780 (6)	627 (6)
No Difference	1448 (7)	1097 (7)
High Compatibility	1075 (98)	968 (77)
Ease of Use		
More Difficult	837 (8)	1257 (8)
No Difference	1206 (6)	1151 (10)
Easier	1089 (92)	895 (72)
Effect on Soil Erosion		
Reduces	1091 (104)	963 (83)
No Change	690 (1)	--- (0)
Increase	993 (6)	865 (7)

[1] Number of respondents is indicated within ( ).

How easy is conservation tillage to use? Although the majority of operators felt it was easier than conventional tillage, actual residue levels did not correspond to this perception. In the case of corn residue, those maintaining the higher residue levels felt there was not a difference between conventional and conservation tillage. The higher levels of soybean residue were being maintained by those who felt they had to work harder at maintaining these levels.

The fact that conservation tillage will reduce soil erosion was accepted by most operators. Yet there was also recognition by a minority that if residue levels were not maintained, then erosion rates will either stay the same or actually increase. In terms of actual numbers, however, there is little recognition that how a conservation tillage system is employed will influence erosion rates.

A complete shift from conventional to conservation tillage involves recognizing, addressing, and working at overcoming a number of obstacles. Providing farmers with glib generalizations on the benefits of conservation tillage does not help them through this process. Instead, they need sound agronomic and economic information to assist them in the adoption process.

Information Sources for the Conservation Tillage Decision

The farmer does not reside in an informational vacuum relative to agricultural knowledge and information. Just the opposite is true. Proponents of conservation tillage must compete with many well-financed distributors of agricultural information. Therefore, who the farmer turns to for what information becomes an important issue for understanding the adoption of conservation tillage (Nowak, 1983a).

Respondents were asked to identify their most useful source of conservation information. As can be seen in Table 4, the majority selected government agencies (e.g., Soil and Water Conservation District representatives, Soil Conservation Service employees, Agricultural Stabilization and Conservation Service employees or county committee members, and county Cooperative Extension Service personnel) followed by personal sources (self, neighbors, friends, and relatives). These sources were also associated with those farmers maintaining the highest levels of corn residue. The highest levels of soybean residue were associated with commercial and university sources.

Respondents were also asked what sources of information they used when actually using conservation tillage for the first time. Note that government agencies now drop to a distant fourth for both corn and soybean residue. Personal sources, although listed by the majority, now rank third in residue levels. However, it is the shift in the residue levels between these responses that is important for understanding how the conservation tillage decision is being made. The highest residue levels are being maintained by those who bypass local sources of information while going directly to the experts. Commercial representatives, which in most cases meant chemical dealers, and university researchers or state extension specialists provided the detailed information these farmers needed to maintain the higher residue levels. On the other end of the spectrum, those individuals relying on the mass media for their information maintained the lowest residue levels.

Table 4. Sources of Conservation Information by Residue Levels

Source	Corn Residue (kg/ha)	Soybean Residue (kg/ha)
Most Useful Source		
Personal	1045 (31)[1]	1004 (26)
Commercial	929 (3)	1070 (2)
University	938 (26)	1089 (21)
Gov't Agencies	1011 (47)	864 (38)
Mass Media	812 (20)	771 (18)
Source Used When First Implementing Conservation Tillage		
Personal	892 (57)	872 (49)
Commercial	1355 (14)	1089 (12)
University	1349 (12)	920 (9)
Gov't Agencies	887 (10)	778 (6)
Mass Media	421 (4)	418 (3)

[1] Number of respondents are indicated within ( ).

Overall, these results are consistent with adoption theory (Nowak, 1983b). Briefly, farmers go through a series of stages in making the conservation tillage decision. Following awareness of the technology they begin to gather information while evaluating its merits relative to their own operation. Often this information is obtained from an impersonal source, but the reliability and validity is then debated within personal circles. If the information they receive is judged to be favorable, then they will try conservation tillage on a small scale basis. Again, they rely on different information sources for any needed assistance. If the results of this trial are satisfactory, based on the evaluation criteria they receive from their particular information sources, then they will move on to full scale adoption.

Another issue concerns how to deliver or present information to assist farmers through the conservation tillage decision. Of the many marketing or distribution techniques available, the effectiveness of two popular methods are presented in Table 5. Respondents were questioned as to whether they attended an agricultural course, clinic, field day, or demonstration in the past year. It was

Table 5. Conservation Courses and Field Days by Residue Level

Attendance	Corn Residue (kg/ha)	Soybean Residue (kg/ha)
Of Those Attending an Agricultural Course or Clinic in the Last Year:		
Related to Conservation	1078 (38)[1]	1088 (27)
Not Related to Conservation	754 (17)	1031 (13)
Of Those Attending a Field Day or Demonstration in the Last Year:		
Related to Conservation	1197 (53)	1164 (47)
Not Related to Conservation	771 (38)	753 (34)

[1] Number of respondents is indicated within ( ).

also determined if the event was related to conservation. Although causality cannot be established on the basis of this data, there is a strong relationship between attending a conservation event and maintaining higher levels of crop residue.

Adapting to a Conservation Tillage System

Adoption involves adaptation. Although one may adopt conservation tillage implements, there is still the need to adapt the implements and one's managerial program to a new set of circumstances. Success in this process, if success is defined in terms of residue levels, should be related to managerial abilities. Table 6 presents one indicator of this issue. The assumption that there is a direct relationship between education and managerial ability is supported by the data in Table 6. This is especially true relative to corn residue, where the management of conservation tillage is more complex. Yet one can only speculate as to the cause for the curvilinear relationship between soybean residue and education.

A final dimension of adapting a conservation tillage system is time. Each crop cycle the farmer learns something new as the tillage system must be modified in response to climate and pest cycles, machinery wear and tear, and new chemical or seed inputs. Therefore one would expect

Table 6. Education by Residue Level

Education	Corn Residue (kg/ha)	Soybean Residue (kg/ha)
Less than High School	810 (20)[1]	735 (20)
High School	911 (77)	902 (63)
Some College	929 (22)	1114 (15)
College Degree	1330 (15)	837 (11)

[1] Number of respondents is indicated within ( ).

that the time one has been using a conservation tillage system would be related directly to one's ability to maintain higher levels of crop residue. This expectation is confirmed in Table 7 where we find those who were among the last to adopt with the lowest average residue levels, while those who were among the first to adopt with the highest average residue levels.

CONCLUSION

The adoption of conservation tillage is a process. It is a process in which the operator's abilities, resources, and institutional position all come into play in adapting to changing agronomic, engineering, and ecological constraints. Most importantly, it is a process that can be either speeded up or hindered through various policy and program considerations.

Research has pointed out that most farmers are adopting conservation tillage systems because of economic objectives (Conservation Tillage Information Center, 1984). Conservation

Table 7. Year Adopted Conservation Tillage by Residue Levels

Year Adopted	Corn Residue (kg/ha)	Soybean Residue (kg/ha)
1977–1980	839 (22)[1]	891 (18)
1973–1976	1100 (45)	983 (33)
1969–1972	1228 (27)	974 (26)
1964–1968	1247 (8)	1269 (5)

[1] Number of respondents is indicated within ( ).

objectives are a fortunate, secondary benefit of this decision. However, what is often ignored in the studies focusing on the diffusion of conservation tillage is the extent to which these conservation objectives are actually achieved. It does little good to point out what is theoretically possible if this cannot be translated into practical results. Recommendations based on modeling efforts have little meaning to the farmer who views the managerial inputs to the model as unrealistic and unworkable. In essence there is a need for researchers and policy makers to treat the result of the adoption decision as a variable rather than as a fixed outcome. This understanding will have immediate implications for our current policy efforts to promote conservation tillage.

Cost sharing is often pointed out as a technique to promote the adoption of conservation tillage. In this capacity cost sharing is used to subsidize initial investment and learning costs. Simply put, cost sharing has been used to finance the risk the farmer must take in shifting to a conservation tillage system. Yet is this the most cost effective approach? Should our policy initiatives attempt to subsidize or minimize risk? These two approaches, subsidizing or minimizing risk, are not the same. In the first case we recognize that the farmer must deal with uncertainties or the lack of adequate information in adopting a conservation tillage system. Because the farmer does not have all the answers to the agronomic and economic consequences of the decision, we pay him to take a risk. In the second case, rather than paying him to take a chance, we try and provide the necessary information to reduce the amount of risk involved.

Up to now we have relied mainly on the first strategy, that is subsidizing risk. If we decide to move toward the second strategy, minimizing risk, then several actions will have to be taken. Basic research on conservation tillage will have to continue. However, more emphasis will have to be put on translating and distributing this information to farmers. In particular, three strategies need to be explored.

(1)   We know that farmers use neighbors, friends, or family to either obtain or assess conservation tillage information. There is a need for more efficient use of these indigenous information networks. These local resources could be used to promote conservation tillage in many ways besides testimonials and sites for field days (Nowak, 1982).

(2)   The Soil Conservation Service and the Soil and Water Conservation District have a number of highly trained technicians to promote the adoption of conservation tillage systems. Yet although trained in agronomic, engineering, and organizational skills, they often lack an understanding of the social and economic theories which govern the local agricultural communities in which they work. A basic understanding of the processes which influence the adoption and diffusion of new technologies would be essential.

(3)   Our limited conservation resources are now being targeted to those areas with the greatest need. Cost sharing for conservation tillage also needs to be targeted, but not necessarily on the basis of erosion rates, threats to productivity, or off-site damages. Instead, these funds should be targeted to those situations where risk cannot be minimized through information assistance strategies. Examples are, supplying machinery to districts where necessary equipment is either unavailable or too expensive, a form of crop insurance for situations where unusual pest circumstances reduce yields when utilizing conservation tillage systems, and as a final example, subsidies to universities, Soil and Water Conservation Districts, or conservation organizations (e.g., Conservation Tillage Information Center) to target information assistance programs to specific areas.

Conservation tillage is revolutionary in nature, but evolutionary in scope. Although a majority of farmers may be currently attempting to use these systems, they are doing so with varying degrees of success. Perhaps we would be better able to help the farmer through this decision process if we stopped counting just the attempts at adoption, and began to pay more attention to the actual degree of success achieved. Then instead of inflated estimates we would have a better understanding of the obstacles the farmer faces when adopting a conservation tillage system. This better understanding will, in turn, provide better assistance.

SUMMARY

The decision to adopt a conservation tillage system is explored based on research results from 193 farmers in Iowa. The adoption process is outlined including major obstacles the farmer must overcome if he is to achieve economic and conservation benefits. Results indicate that a number of farmers who have adopted the implements of conservation tillage have not developed the necessary supporting managerial skills. This is investigated by looking at perceptions of conservation tillage as well as sources of information and assistance. A number of recommendations are made concerning how to promote conservation tillage systems.

LITERATURE CITED

Block, J.R. 1983. The national program for soil and water conservation. Testimony before the U.S. Senate Subcommittee on Soil and Water Conservation, Forestry, and Environment. March 9. Washington, D.C.

Conservation Tillage Information Center. 1983. 1982 National Survey Conservation Tillage Practices. Ft. Wayne, Indiana.

Conservation Tillage Information Center. 1984. A Survey of America's Conservation Districts. Ft. Wayne, Indiana.

Crosson, P. 1981. Conservation and Conventional Tillage: a Comparative Assessment. Soil Conservation Society of America. Ankeny, Iowa.

Crosson, W.D. 1982. Conservation tillage: an available solution. Presentation to the University of Florida Humanities and Agricultural Program Conference. October 19.

Hinkle, M.K. 1983. Problems with conservation tillage. J. Soil Water Cons. 38(3):201-206.

Kinney, T.D. 1983. Agricultural Research. March. U.S.D.A. Washington, D.C.

Lessiter, F. 1974. 1973-1974 no-till farmer acreage survey. No-Till Farmer (March):6.

Lessiter, F. 1983. 1982-1983 no-till farmer acreage survey. No-Till Farmer (March):9.

Nowak, P.J. 1982. Strategies for the conservation district. Unpublished workshop materials. 2305 E. 5th St., Duluth, Minnesota.

Nowak, P.J. 1983a. Obstacles to the adoption of conservation tillage. J. Soil Water Cons. 38(3):162-165.

Nowak, P.J. 1983b. Adoption and diffusion of soil and water conservation practices. Rural Sociologist 3(2):83-91.

Nowak, P.J. and P.F. Korsching. 1983. Social and institutional factors affecting the adoption and maintenance of agricultural BMPs. F.W. Schaller and G.W. Bailey (eds.). Agricultural Management and Water Qaulity. p. 349-373. ISU Press, Ames, Iowa.

Nowak, P.J. and P.F. Korsching. 1984. Sociological factors in the adoption of best management practices: a final report. U.S. Environmental Protection Agency, Athens, GA.

Office of Technology Assessment, Congress of the U.S. 1982. Impacts of Technology on U.S. Cropland and Rangeland Productivity. Washington, D.C.

Pioneer Hi-Bred International, Inc. 1982. Soil Conservation Attitudes and Practices: The Present and the Future. Des Moines, Iowa.

U.S. Department of Agriculture. 1975. Minimum Tillage: A Preliminary Technology Assessment. Washington, D.C.

CHAPTER 25

PUBLIC POLICY ISSUES INFLUENCING DIRECTIONS
IN CONSERVATION TILLAGE

Lawrence W. Libby
Department of Agricultural Economics
Michigan State University
East Lansing, Michigan 48824

INTRODUCTION

Decisions on content in a paper at the end of a
major conference such as this must be made partly on the
basis of what may have been left unsaid. It is hazardous
to predict, but one must do so. Because the conference
theme has been carved into many pieces already, my purpose
is to add a policy perspective to technical content presented
by others. I will focus on the utility of all that technical
information in making public choices among ways to influence
private choices in adoption of conservation tillage.

Soil conservation policy, including policy for conservation
tillage, includes a broad range of social, political and
economic techniques that influence the rights, obligations
and opportunities of farmers whose actions affect soil
erosion. The term "policy" refers to the flow of those
factors, and does not imply consensus on a given set of
actions. Manipulating the rights, opportunities and obligations
facing the farmer will presumably change the private choices
made.

The farmer must constantly make decisions in running
a farm. His or her options are constrained by property
rights and the rights of others as defined in law and
practice. Each legal option in a particular decision
carries certain predictable consequences, including impacts
on parties outside the farm enterprise. Farmers are reasonably
rational, like the rest of us. They make tillage choices

based on likely impact on their ability to earn a living, maintain quality of land as an economic asset, keep peace with their neighbors, and keep peace with their own conscience. Land stewardship is indeed one of several motives affecting a farmer's land use decisions. Production residue such as sediment, nutrients, pesticides and pathogens may create substantial problems for landowners and water users well beyond the farm boundaries.

This paper adds a policy perspective to selected sets of facts about conservation tillage. The following questions are addressed:

(1) Why should governments at any level be tampering with the farmer's choices in selection of a tillage system? We know they are--many of the "tamperers" are with us at this conference. But there are many other things governments could be doing, perhaps with greater overall payoff to society. While some may argue that it is hopelessly academic (or futile) to infer a logic in government actions to reduce soil erosion (Sampson, 1981), I disagree. There has to be something more than inertia in public budgeting. As the political setting for soil conservation policy becomes more complex (Leman, 1982; Libby, 1982a), the need for clear and defensible rationale for spending public funds on soil conservation becomes more pronounced.

(2) What is the policy relevance of the physical and economic consequences of differenct tillage systems, within a given structure of property rights?

(3) How may the "right to erode" be reallocated to change the farmer's choice of tillage system?

(4) What are intra- and inter-governmental implications, including bi-national concerns with Canada?

Each of these questions will be addressed in a somewhat cursory manner, given constraints on time and printing space; however, an attempt will be made to leave a trail of citations for anyone interested in pursuing a particular point.

WHY PUBLIC POLICY FOR REDUCED TILLAGE?

The standard economic logic in this question is that government tampers in a person's life only when the social consequences of what that person does or fails to do exceed private consequences of the choice by an amount sufficient to generate political demand for action. Government does not sit around looking for things to do,

but when those bearing social impact have sufficient clout
to get attention, government responds. Many economists
argue that government should act only when there are these
measurable non-pecuniary externalities.

If we accept this general notion that government
is supposed to tamper only when the private decisions
in question create benefits or impose costs on people
who have no part in the decision, why should government
be pushing conservation tillage? As we have heard at
this conference, soil erosion can reduce soil productivity
on the farm and create water pollution and silting problems
off the farm. The latter situation is more clearly an
economically defensible rationale for government tampering
than is the former. Water pollution is the standard example
of negative externality where private actions impose real
costs on others. Yet water pollution has not been given
high priority in the Secretary of Agriculture's preferred
soil conservation program submitted to the President and
Congress last year (USDA, 1982). The U.S. Environmental
Protection Agency (EPA) and Department of Agriculture
(USDA) are combining their efforts to deal with this off-farm
result of erosion (Non-Point Source Task Force, 1983).

Most of the discussion at this conference has focused
on the on-farm aspect of erosion and use of conservation
tillage to reduce that erosion. Often, the same farm
level choices and conservation techniques have impacts
both on and off the farm. The policy question is if our
primary concern is soil productivity, and if conservation
tillage makes a difference, what are the social advantages
to conservation tillage that make it reasonable for government
to try to bribe, force or beg farmers to use it? Does
reduced tillage create a benefit to society beyond that
realized by the farmer? In the absence of government,
why are we under-investing in conservation tillage? Or
why are we under-investing in the soil component of production
functions for food?

Pierre Crosson at Resources for the Future has done
the most perceptive writing in this topic (Crosson, 1982;
1983). He sees two possible sources of market failure:
(1) the market does not adequately record the returns
to conservation investments that will maintain soil productivity;
or (2) even if the market works, various institutional
factors limit the farmer's ability to respond to market
signals that soil conservation is warranted. To support
the first argument, we must assume that farmers lack economic
information on productivity effects of erosion, and thus
the payoff from reduced tillage, because erosion is difficult
to detect. Perhaps they lack information on future food

demands that suggest economic returns to soil productivity, or their economic planning horizon is too short and discount rate on future farm income too high, or they over-estimate availability of future soil-replacing technologies. The market for farmland productivity and conservation investment is the result of collective judgments on these factors. Given that all of these assumptions are accurate, we still must assume that government is better able to cope with these information gaps than is the farmer whose livelihood depends on the result. Crosson correctly points out that the time frame/discount rate argument is specious, because land is a capital asset for the farmer, with a present value determined by discounted returns to land in perpetuity. The land market does not conform to the income-earning life of a particular manager. "Regardless of his time horizon, therefore, he has the incentive to control erosion so long as the present value of the cost of control is less than the capital loss" (Crosson, 1984). That general rule applies to government as well. If the present value of public expenditure on soil conservation exceeds the present value of returns from protecting soil productivity, then we are saving more soil than we need and in the process may be imposing inefficiencies on future generations of producers and consumers. Basically, the farmer has a real incentive to know about the economic effects of erosion on productivity and therefore the payoff to investment in soil conservation practices, including reduced tillage. A farmer will pay less for eroded or erosive land than for other land. He or she is implicitly projecting the economic importance of soil in production functions of the future. Is government better able to do that than the farmer? Perhaps the most defensible public investment is in information, to help the farmer make rational choices about soil conservation.

On the second line of argument, perhaps institutional factors create incentives that mask or distort market signals, thus leading to less private investment in conservation tillage that is socially optimal. Contrary to the popular presumption that tenants care less about soil conservation than owners, recent findings by Linda Lee at Oklahoma State University indicate that a smaller percentage of owner-operators than either part owners or landlords have adopted conservation tillage (Lee, 1983). Absentee owners have the same incentives to compare conservation investment to erosion cost as do owner-operators.

Evidence of government-induced erosion may be the most compelling rationale for public policy, if only to control these unintended side effects of policy designed for other purposes. Various programs are undertaken to improve incomes for farmers by improving price stability or shifting some of the risk from individuals to the public. Many of these create implicit incentives to till erosive land. If the farmer is insulated from the full impact of misuse of resources, private incentives for land stewardship may be masked. Crop insurance, disaster payments and even price supports are designed to cushion the economic effects of various unfortunate events, many of which are beyond the direct control of the producer. But disaster funds tend to get widely distributed--in recent years nearly two-thirds of all counties in the U.S. were eligible for disaster funds. All of these programs sustain production levels higher than warranted by market conditions.

High production may be encouraged to meet export levels desired for various national policy reasons. Export policy has encouraged major increases in use of erosive land. The farmer is simply responding to incentives and opportunities put before him by government action. The most erosive land goes in and out of production as conditions change from year to year. One writer observed that "in one year, we may have lost the equivalent of soil saved by the Agricultural Conservation Program in three to five years at a cost of $238 to $247 million" (Cook, 1983).

Similarly, a tax policy that taxes capital gains at a lower rate than income encourages farmers to buy land, thus shifting income to capital. It is cheaper to buy land than to adequately protect land already owned. Depreciation allowances on irrigation equipment and land conversion machinery encourage farmers to bring new land into production, land that might not be productive without those massive infusions of capital (Raul, 1980). Thus, because government creates erosion, perhaps government should take corrective actions to reduce it. Consistency among these bundles of incentives offered to farmers to alter their behavior in prescribed ways is being examined by the Economic Research Service of USDA.

The market failure line of argument assumes that government's role is to root out "inefficiency" wherever it occurs. In a political sense, though, we know that government responds to demands for selected rule changes to produce a different mix of goods and services simply because people with clout want it that way. The political environment for soil conservation is far more diverse now than it was pre-RCA. Many national environmental groups have soil conservation policy positions. There are political demands that farmers be asked to do things differently, to produce a different combination of food, soil productivity and water pollution than is currently produced. Efficiency is involved only as a set of standards to guide public spending in ways that can produce greatest prossible erosion-reduction for the cost.

Much of the political demand for soil conservation programs is based on the judgment that we should be careful with the scarce supply of soil productivity, to allocate its use in such a way as to assure that future generations have adequate food supply. Perhaps there will be adequate soil replacing technology to continue recent increases in output per acre. Perhaps farmers are indeed in the best position and have the greatest incentive to protect land quality. Perhaps land markets will signal scarcity far more accurately than goverment can. But a key question is, should we take the chance? Markets are but one means for allocating risk, based on the consequences and probabilities of various outcomes. Perhaps the social consequence of under-investing in soil productivity is greater than the sum of consequences for individual investors, and the role of government in this case is to be sufficiently cautious to avoid the catastrophic result of providing too little soil productivity for the future. The cost of buying more soil conservation than needed is the value of goods and services we as a society could have acquired with those funds. I would argue that we are better off with the cost of too much soil productivity than of too little (Little, 1983).

THE POLICY RELEVANCE OF ALTERNATIVE TILLAGE SYSTEMS

The second major policy perspective on conservation tillage involves the impacts of these different tillage systems for the farmer. Most of the papers at this conference involve specifying the technology of reduced tillage. This information has value to the extent that it helps the farmer anticipate consequences of reduced tillage within the rights and opportunities he currently has available. Farmers are being told that they need not sacrifice income while conserving soil to improve water quality and assure production capacity for future generations. Reduced tillage, we have heard, offers the best possibility of achieving socially desirable levels of soil conservation without the attendant private costs of the structural options like terraces or diversions. For the same price, the farmer can produce food and protect soil, rather than producing food plus erosion. Thus, our engineering and agronomic research has utility by helping the farmer understand what will happen to or for him if he selects one tillage system over another. The decision is left to him. There are several important policy aspects.

(1)   Given that the farmer has discretion in this matter (he can change to reduce tillage or not) some effort must be made to understand his motives. That is, if farmers can "take it or leave it," and taking it has social value, what sorts of incentives guide farmer choices? Understanding and predicting farmer behavior would have real utility in policy.

Traditional economics suggest that farmers act as if they were profit maximizers. The assumption that farmers will make decisions that increase income or asset value is fairly useful in predicting how farmers will respond to such economic signals as cost or price changes. Many analyses and policy prescriptions are built on that assumption. They seek to determine optimal investment in conservation by farmers operating with varying combinations of soil, cropping patterns, and prices. Eleveld and Halcrow point out that there is a difference between level of conservation that is rational for the farmer and for society. The message is that since society wants more conservation than the individual producer considers to be economically rational, then society must pay the extra cost (Eleveld and Halcrow, 1982). Pope et al. (1983) have found that for most Iowa soils, conservation tillage can reduce erosion with less reduction in farm profits than is true for other conservation techniques. Oscar Burt points out (Burt, 1981) that farmers in the Palouse area of Washington find continuous wheat production to be economically rational in the short run and the long run, even though crop rotation is less erosive. Again, there is an implicit policy prescription here--that the economic returns to farmers must be adjusted if we are to expect soil conserving effort. Emphasis on these and similar studies is on the economics of production under varying tillage and cropping patterns, not on policy.

There have been recent efforts to focus directly on policy by seeking a more complete understanding of why farmers do what they do. Armed with this understanding, the policy makers can adjust various incentives to encourage conservation. In a recent study for EPA, Miranowski et al. (1983) developed and applied a probability model for choice of a tillage system as a function of various characteristics of the farm operator. Their conclusions suggest who conserves, though do not directly question why. Farmers with more education, those on smaller farms, and those on hilly, erosive terrain were most likely to adopt a reduced tillage system. Tenure of the operator was not an important factor. In a current study at Michigan State University, Procter is seeking a more accurate theoretical formulation of farmer behavior with respect to a conservation decision (Procter, personal communication). There is a need for

more empirical work to further clarify behavioral variables
relevant to the soil conservation decision. Results will
suggest policy opportunities within the current allocation
of property rights that leaves the right to erode with
the farmer. To the extent that erosion creates off-farm
costs in the form of pollutions some of that right to
erode has already been recalled by the government.

(2)   Consequences of alternative tillage systems
must be tied more specifically to varying terrain, soils
and cropping sequences. If we are going to give the farmer
complete freedom in tillage decisions and rely on better
information of economic impact to sell reduced tillage,
then that impact information must be tailored to differences
among farms. To paraphrase Abraham Lincoln, you can fool
some of the farmers all of the time, and all of the farmers
some of the time, but you can't fool all of the farmers
all of the time. And I contend that farmers are harder
to fool than most people. In our enthusiasm for conservation
tillage, we simply have not been candid enough about differences
of impact. The general effects of alternative conservation
tillage systems have been well documented (Mueller et
al., 1981; Phillips et al., 1980). We know that reduced
tillage retains soil moisture, thus often improving yields
on droughty soils, particularly in dry years. This may
be a distinct disadvantage on poorly drained soils. Reduced
tillage tends to retard the rate of soil warming which
may delay germination in some soils, in some years, in
most northerly areas. Reduced tillage may reduce effectiveness
of pre-emerge and water soluble herbicides and in other
ways increase annual and perennial weed problems. This
is more of a problem on some soils than others. Similar
problems are encountered in insect and disease control
under reduced tillage, because there is greater surface
residue to transmit disease. Nitrogen availability is
reduced on some soils with conservation tillage systems,
while phosphorus is apparently more effective. Conservation
tillage usually involves significant reductions in fuel
requirements, though for some farms increased cost of
herbicides and other energy intensive inputs more than
offset fuel savings (Lockeretz, 1983).

(3)   The off-site impacts of alternative tillage
systems must be documented more accurately. If we are
to make the case that government should be subsidizing
use of reduced tillage because the social gains are greater
than private gains, then it seems reasonable that net
social impact should be estimated, including impacts off
the farm. EPA is promoting conservation tillage as an
effective practice to reduce non-point pollution. The
basic reasoning is that if conservation tillage reduces

erosion and is economically attractive for the farmer, then it must be the best way to reduce the off-farm impacts of erosion as well. There has been little empirical work, though, on the negative off-farm consequences of reduced tillage. In previous writing (Baker and Laflen, 1983) and at this conference, James Baker has very effectively discussed linkages between tillage systems and chemical concentrations of run-off water. While no-till may reduce run-off, attendant increases in pesticide use may increase chemical concentrations in water that does leave the field. Baker reviews several field and experimental studies that document that general conclusion. He also suggests techniques for reducing this chemical displacement problem.

Maureen Hinkle has also suggested potential problems inherent in substitution of chemical pesticides for fuel in moving to a reduced tillage system (Hinkle, 1983). The key point is that there are entries on the negative side of the ledger that must be acknowledged, analyzed and communicated to the farm community and to policy makers. Reliance on the education or "jawboning" approach to soil conservation policy in which farmers retain all rights currently available, including the right to erode, requires that impact information be complete and specific.

CONSIDERING ALTERNATIVE WAYS TO CREATE AN OBLIGATION TO CONSERVE

The inherent appeal of conservation tillage is that under many circumstances the farmer can serve his own interests and those of society at the same time. This is a rare situation in policy and should not be taken lightly. Any policy option implies an initial distribution of property rights that establish the context for choice. Any change in policy carries an adjustment in incentives for action in the system, with some participants able to shift the inevitable cost of that change onto others. So far, land owners have been able to shift most of the cost of soil conservation to taxpayers by retaining full discretion in land use. Farmers may bear a cost, but only if they want to. And many farmers have accepted responsibility for soil stewardship, practicing soil conservation when not clearly in their economic interest.

The glossary of national soil conservation policy has, until recently, been completely free of words that imply mandatory action by the land user. Even the Environmental Protection Agency, that venerated regulator of polluters, has studiously avoided any suggestion that the right to permit non-point pollution should be recalled by the public. Conservation tillage has been a cornerstone of non-point abatement policy in the Great Lakes Basin because it "...can have a major impact on Great Lakes water quality improvement with minimal effect on agricultural profitability" (Non-Point Source Task Force, 1983). Emphasis has been placed on technical assistance and demonstration. While I would not argue here that all farmers should be required to conserve soil or that mandatory measures should be emphasized, it is important that these options be retained on the policy agenda. Reordering the rights and obligations of farmers in the interest of reducing erosion, run-off and the multi-media transport of pollutants is both legally and politically acceptable (Libby, 1982b). Land owners' choices are redefined for other purposes--like farmland preservation or rural land use planning--why not for soil erosion? Selling reduced tillage as a conservation strategy must be analyzed as but one point on the spectrum of allocation of rights between owner and public. If our only choices are voluntary measures for reducing erosion, conservation tillage is bound to be extremely attractive. But, options that redistribute rights must be compared as well, in an open and careful manner. When costs of implementation are added to direct outlay of public cost for technical assistance, it could be that softer voluntary approaches still come out ahead.

More mandatory approaches come in several forms:

(1)   Outright regulation against erosion exceeding a pre-determined limit has been tried in several states and localities. A model sediment control ordinance was developed in 1972 and circulated among the states (Libby, 1982b). Only four states, Iowa, Michigan, Pennsylvania, and South Dakota, have ordinances that include agriculture. Agriculture was added to the Michigan law as something of an afterthought to avoid the possibility of more stringent regulation from the federal level under the 208 Water Quality Planning program. The Iowa law is widely viewed as the most aggressive. It established soil loss limits for different soils and requires that farmers comply if there is cost sharing available.

County soil conservation districts have the authority to regulate soil loss in most states, though few have exercised it. There are exceptions like the widely reviewed Sterling Township in Vernon City, Wisconsin, which requires good conservation methods on land of greater than 6 percent slope (Barrows, 1983).

(2)   Enforcement by humiliation seems to be the strategy employed in the recently enacted state program in Illinois. The state has developed erosion control

guidelines based on T value of Illinois cropland, and passes those guides along to the county soil conservation districts for conversion to standards. County standards, required to be at least as stringent as state guidelines, establish erosion limits that phase in over several years to achieve "T by 2000"--the theme. The enforcement process begins with a formal complaint filed by an individual or group, including the soil conservation district. The district then works with the offending farmer, providing technical assistance and encouragement, to bring his compliance up to standard. The relucant farmer must participate in a local public hearing to justify or explain his failure to comply. The record and findings of the hearing are made public. If he still doesn't respond, a higher order of public humiliation is imposed--a state level public hearing. In front of his friends, family and peers, presumably with the whole world watching via satellite television, the farmer must explain why he has failed to heed the cries of future generations and turned his back on stewardship. A fine would be a welcome alternative to such an ordeal (Walker, 1982).

(3) Various cross compliance measures represent the softer set of mandatory soil conservation options. The farmer's set of opportunities are redefined to link certain obligations to options currently available. Choice remains with the producer, though the incentives are altered to produce a different set of consequences associated with each option. In Wisconsin, for example, a farmer is eligible for the special farmland preservation tax incentive only if he conducts farming in accordance with an approved soil conservation plan. Michigan's Farmland and Open Space Preservation Act (PA-116) is very similar to the Wisconsin tax program, but without the soil conservation requirement. Perhaps depreciation allowances for irrigation and other equipment, much of which facilitates bringing new and erosive land into production, should be available only if soil conservation practices (including reduced tillage) are employed. Commodity programs cost taxpayers millions of dollars every year. Perhaps taxpayers should get some return on their investment in the form of conservation and tillage practices that are sensitive to offsite effects and the public stake in protecting soil productivity.

Cross compliance simply raises the potential cost of failing to conserve while leaving the choice with farmers. Reduced tillage would likely gain in popularity under this approach, as the least costly way for a farmer to retain eligibility for positive incentive programs currently available.

(4)   A <u>cost effectiveness approach</u> to achieving soil conservation standards would likely push the farmer toward the least costly technology.[1]   The suggestion here is that SCS and ASCS determine the least costly set of structural and non-structural measures to achieve acceptable erosion levels and provide cost-sharing and technical assistance <u>only</u> to the level necessary to get those particular measures in place.   The farmer may choose a different approach, but will do so at his own expense.   The general notion is that taxpayers need not pamper the farmer who avoids reduced tillage simply because he doesn't like it, or it is a departure from family tradition.   There is no reason for taxpayers to subsidize such stubbornness by continuing to provide assistance for some less efficient set of practices.

(5)   <u>Variable cost sharing and targeting</u> represent a more "active" approach to distributing public funds than has been practiced in the past.   Presumably, society achieves greater conservation at the same public price. To the extent this "bang for the buck" approach to allocating funds redistributes effort, some people are going to lose. In fact, if no one loses the change probably has very little substance.   While there is intuitive appeal to more efficient use of public funds, efficiency carries substantial political baggage and those whose interests are damaged by the change have their own chain of logic that may be just as appealing.   Neil Sampson, former Executive Director of the National Association of Conservation Districts (Sampson, 1982), has argued,

> ....soil conservation is at its heart a constant core maintenance program for the land, it is both unwise and destructive to turn the program on and off because of perceived 'breakouts' of the problem in certain areas under certain conditions....Targeting works well for moving around inanimate objects, like dollars.   It works much less well and with far more damaging effect to the targeted programs when using 'live' ammunition--trained people.   Those SCS technicians need to gain local credibility and maintain it--you simply can't use federal hit squads to run around the country swatting at farmers who are causing excess soil erosion.   You have to have a stable, credible program...and not go off chasing new hot spots.

---

[1] I am indebted to Tony Grano, Economic Research Service, for these ideas.

Neil always could turn a phrase, and make an astute political point at the same time. Economic efficiency as a set of rules for resource allocation assumes a degree of resource mobility that entails considerable dislocation cost for people. Only those not directly affected really care very much about economic efficiency as a policy goal. Soil conservation districts in areas where conservation maintenance rather than emergency action is the rule of the day are unlikely to be enthusiastic about targeting.

Perhaps it is my imagination but I detect a slight cooling of relationships between SCS and NACD. Their cooperation was essential in framing the Soil and Water Resources Conservation Act (RCA). It is the implementation phase that is giving them problems. Speaking to Soil Conservation Service staff recently, Assistant Secretary of Agriculture John Crowell admonished SCS staff to represent the Administration budget position in this area and not get too cozy with the districts (Crowell, 1983).

> The local districts...are more concerned with getting their fair share of the resources provided by the Federal government than they are in seeking the best nationwide application of those resources for accomplishing the maximum amount of soil protection and water conservation. 'Fair share' to the districts means at least as much as they've been getting from the Federal government in the past, and some part of any new resources.

Crowell expressed the judgment that districts have taken an inordinate role in setting conservation priorities and ordered soil conservation staff to stay clear of district officials at budget time.

Variable cost sharing is an effort by ASCS, working with SCS, to apply a higher cost sharing rate to those areas with greatest erosion. The obvious notion is to provide extra incentive where the payoff would be greatest. Criteria for selection of variable cost share counties have included evidence of local willingness to participate as well as soil loss information. It could hardly be characterized as an experiment in a new delivery technique for soil conservation practices, because there is no overall design, no control and little monitoring data. But it does represent a structured pilot attempt to shift effort toward those areas having greatest erosion to tempt the rational economic eroder to behave in a socially responsible manner. Measures of performance are largely physical at this point--reduced erosion, measured in tons of soil per acre per year for the cost-share dollar spent. In addition, some indicator of the relative significance of a ton of soil movement under differing soil and cropping conditions would be valuable. A ton is not necessarily just a ton--some are worth more than others. Reducing soil loss on a productive though shallow soil could have greater impact on productivity in the short or immediate term than a ton of deeper soil. Priorities relative to long term depletion are more difficult. Value of that productivity loss would differ depending on the crop. These aren't the only decision criteria, but economic performance information should play some role in targeting of conservation funds (Kugler, 1984)

INTER- AND INTRA-GOVERNMENTAL POLICY ISSUES

A major tenet of soil conservation policy in the U.S. has been its intergovernmental character. Local units, particularly soil conservation districts and ASC county committees are closest to the farmers, sensitive to realities of production. Delivery of soil conservation practices and knowledge have been a local function. The Feds contribute the disciplinary expertise--technical standards, support for soils and engineering specialists who can put practices in place. States have functioned primarily as a pass-through level, with a charge to "bless" the funds and staff priorities as they filter out to locals. State level sediment control programs have relied on the SCD's for implementation and for local support. The governmental balance of power in soil conservation policy has definitely been someplace very close to the county. The structure is an impressive one, somewhat the envy of other resource agencies like EPA and the Corps of Engineers, both of which are now talking about soft approaches to policy through the local units. There has been a considerable amount of local experimentation in delivery of conservation tillage technologies. Conservation districts and county committees have developed various imaginative ways to encourage adoption of reduced tillage, including special leasing and sharing arrangements for tillage equipment. No one really knows the range of these special incentive arrangements being tried--the information should be collected.

The RCA program marks a potential realignment among levels of government, a possibility recognized by Crowell in his speech to Soil Conservation Service professionals, noted above. RCA is but one manifestation of the growing public demand for greater accountability in natural resource policy. There are many socially defensible uses for dollars currently invested in natural resources. Prior to RCA, there had been demands for evidence of impact from federal dollars spent for soil conservation (Leman, 1982). Greater attention to a systematic appraisal of soil conservation needs and pay-off inevitably means greater centralization in the system. The analytical back-up for RCA places a premium on those professionals who can manage large information systems--a centralized activity. The concept of geographic targeting implies some centralized judgment as to who will gain at whose expense. The key policy issue here is whose targets are to guide conservation effort. The RCA law itself definitely implies national priorities. Early drafts of the Secretary's Preferred Program proposed to rely entirely on state level priorities for targeting on the implicit assumption that the sum of state priorities makes sense for the nation as a whole. National targets would mean that some states would lose (likely including Michigan); state level targets mean that some districts would lose (SCS support for several districts in urban counties in Michigan has already been eliminated or sharply cut); district level targets would mean that everyone would lose (through inefficient use of funds) though nobody would lose very much.

While there will be some strain on the system, I feel that national soil conservation programs will continue to centralize somewhat. Soil Conservation Districts may well go their own way, seeking funds from the states, EPA and other sources as needed to conduct soil and water programs of local interest. Relationships between NACD and the federal establishment will get increasingly testy at the national level, though will endure a bit longer in the states.

Within the federal level of government, USDA and EPA have essentially divided the soil conservation turf between on-farm productivity concerns and off-farm water quality issues. There are special water quality funds available from EPA to encourage adoption of reduced tillage in special project areas. While this lack of coherence in Washington-level attention to erosion is of concern to some, particularly local units trying to cope with administrative confusion, there may be utility in "cooperative competition" among the agencies. Each has strengths and limitations based on their historic missions. USDA has always been the friend of the farmer, providing help and advice when desired, but has lacked the "killer instinct" necessary to push a particular action to completion. EPA has had plenty of experience in forcing issues, but has lacked the soft human touch with people whose actions ultimately cause the problem. Together they may be a great team.

The greatest organizational need in soil conservation policy, including conservation tillage, is a better mechanism for cooperation with Canada in designing institutions that will protect the quality of the boundary waters and will retain productive agriculture on which both nations depend. The U.S. has gone further than Canada in its structure for conservation policy. Conservation Authorities in Canada might well perform many of the functions of our SCD's, but they do not at this time. The record is, however, that once motivated, Canadian institutions can respond quickly and forcefully. The binational PLUARG effort and subsequent evaluation, conducted under the auspices of the International Joint Commission stand as the best state-of-the-art appraisal of non-point abatement efforts in the two nations (International Joint Commission, 1983). The problem is that we still lack the specific organizational mechanism to focus on soil erosion and particularly on the performance of alternative means for controlling it. The simple process of sharing experiences among local units of government that have enacted rules to control erosion would be of significant value in implementing our good intentions. Perhaps a joint U.S.-Canada panel on soil erosion, to establish an information exchange and an action process for implementation would be valuable, outside of those government agencies with the mission to implement a particular set of policies.

CONCLUSIONS

What does all of this mean for the future of conservation tillage adoption? The following final observations are offered:

(1) The push for accountability in natural resources policy, and particularly the performance emphasis in RCA, will make reduced tillage an increasingly attractive conservation alternative. There will be differences among reduced tillage options for various farms and resource mixes. Studies have shown that under many conditions, reduced tillage is economically rational on its own; in other circumstances it costs no more than conventional tillage systems after initial conversion cost. The related point here is that if reduced tillage pays for the farmer, there is no obvious reason why taxpayers should pick up part of the operating cost, beyond some initial help and continued education. More soils and crop data will be needed to establish the physical consequences for the farmer and data on economic impacts of those physical changes. Further, if conservation tillage is the least cost way to achieve some designated reduction in erosion in a particular farm, there is no reason for taxpayers to pay more than that

minimum amount just because farmers prefer to stick with their old tillage system. Even if farmers retain the right to erode, the rest of us should not have to pay more than the minimum share necessary to put the most efficient technology in place. If farmers continue to abuse the right to erode, more mandatory measures may be necessary.

All of this requires better farm-specific data. Investment in performance information can be money well spent.

(2) More must be done to net out the various on-farm and off-farm results of tillage alternatives. Reduced tillage retains productive soil and reduces runoff to nearby water bodies, but to differing degrees depending on soil type and cropping pattern. While we should not expect an optimal allocation of soil conservation expenditures based on all on- and off-farm effects, we should at least generate field and experimental data that bring the two together. It is conceivable that reduced tillage under some circumstances would produce unacceptable reductions in water quality, imposing clean-up costs on someone else. A more efficient strategy might involve structural measures that cost more to install but because compensating pesticide application is not acquired, will preserve water quality.

(3) So far, there has been inadequate accounting of the various transactions costs involved with implementing reduced tillage. We simply need to keep better track of the kinds of private incentives being introduced by machinery dealers, chemical manufacturers and financial institutions to encourage farmers to shift tillage systems; the degree to which financial attractiveness of these efforts depends on technical assistance, education or cost sharing by government; and the overall investment of technician time and support funds to push reduced tillage.

LITERATURE CITED

Baker, J. and J. Laflen. 1983. Water quality consequences of conservation tillage. J. Soil and Water Cons. 38(3): 186-193.

Barrows, R. and C. Olsen. 1981. Soil conservation policy: Local action and federal alternatives. Am. J. Agr. Econ. 36(6): 312-316.

Burt, O. 1981. Farm level economics of soil conservation in the Palouse Area of the Northwest. Am. J. Agr. Econ. 63(1): 83-92.

Cook. K. 1983. Commentary: Surplus madness. J. Soil and Water Cons. 38(1): 25-28.

Crosson, P. 1982. Diverging interests in soil conservation and water quality: Society vs. the farmer. Perceptions, Attitudes and Risks: Overlooked Variables in Formulating Public Policy in Soil Conservation and Water Quality, L. Christensen and J. Miranowski (eds.), Washington, DC: Economic Research Service, USDA. ERS Staff Report No. AGES820129, pp. 50-69.

Crosson, P. 1983. A perspective on the appropriate role of the public sector dealing with the uncertainties of the impact of soil loss. Perspectives on Vulnerability of U.S. Agriculture to Soil Erosion: An Organized Symposium, L. Christensen (ed.), Washington, DC: Economic Research Service, USDA. Staff Report No. AGES830315, pp. 33-42.

Crosson, P. 1984. Soil erosion and soil conservation policy in the United States. Unpublished draft of material for American Agricultural Economics Association Task Force on Soil Conservation Policy.

Crowell, J. 1983. Looking ahead at SCS--1984 and beyond. Unpublished remarks prepared for delivery by Assistant Secretary of Agriculture John Crowell.

Eleveld, D. and H. Halcrow. 1982. How much soil conservation is optimum for society? Soil Conservation Policies, Institutions and Incentives, edited by Halcrow, Heady, and Cotner, Ankeny, IA: Soil Conservation Society of America, pp. 233-250.

Hinkle, M. 1983. Problems with conservation tillage. J. Soil and Water Cons. 38(3): 201-206.

Kugler, D. 1984. Variable cost-sharing level policy implications for Kentucky's Jackson Purchase Area: An economic study of cash grain production considering soil depletion. Unpublished Ph.D. dissertation. Department of Agricultural Economics, Michigan State University.

Lee, L. 1983. Land tenure and adoption of conservation tillage. J. Soil and Water Cons. 38(3): 163-168.

Leman, C. 1982. Political dilemmas in evaluating soil conservation programs: The RCA process. Soil Conservation Policies, Institutions and Incentives, edited by Halcrow, Heady, and Cotner, Ankeny, IA: Soil Conservation Society of America, pp. 47-88.

Libby, L. 1982a. Controlling soil erosion by rules and regulations. _Proceedings of a Conference on Public Policy Issues and Soil Conservation_. S. Beyer and L.C. Johnson (eds.), Madison WI: Soil Conservation Society of America, Wisconsin Chapter, pp. 41-55.

Libby, L. 1982b. Interaction of RCA with state and local conservation programs. _Soil Conservation Policies, Institutions and Incentives_, edited by Halcrow, Heady, and Cotner, Ankeny, IA: Soil Conservation Society of America, pp. 112-128.

Libby, L. 1983. A perspective that strong public action is needed to deal with the problems of soil erosion. _Perspectives on the Vulnerability of U.S. Agriculture to Soil Erosion: An Organized Symposium_. L. Christensen (ed.), Washington, DC: Economic Research Service, USDA, Staff Report No. AGES830315, pp. 43-53.

Lockeretz, W. 1983. Energy implications of conservation tillage. J. Soil and Water Cons. 38(3): 207-211.

Miranowski, J., M. Monson, J. Shortle, and L. Zinser. 1983. The effect of agricultural land-use policies on stream water quality: Economic analysis. Athens, GA: Environmental Research Laboratory, Environmental Protection Agency.

Mueller, D., T. Daniel, and R. Wendt. 1981. Conservation tillage: Best management practice for non-point runoff. Environmental Management. 5(1): 33-53.

Non-point Source Control Task Force. 1983. Non-point source pollution abatement in the Great Lakes Basin: An overview of post PLUARG developments, Windsor, Ontario: International Joint Commission. Report to the Great Lakes Science Advisory Board, International Joint Commission.

Phillips, R., R. Blevens, G. Thomas, W. Frye, and S. Phillips. 1980. No tillage agriculture. Science. 208: 1108-1113.

Pope, A., S. Bhide, and E. Heady. 1983. Economics of conservation tillage in Iowa. J. Soil and Water Cons. 38(4): 370-373.

Procter, R. "Optimal control theory and the political economy of agricultural non-point pollution control. Ph.D. dissertation underway.

Raup, P. 1980. _The Federal Dynamic in Land Use_. Washington, DC: National Planning Association, NPA Report No. 180.

Sampson, N. 1981. Farmland or Wasteland, Emmaus, Pennsylvania: Rodale Press.

Sampson, N. 1982. Unpublished correspondence.

United States Department of Agriculture. 1981. Program Report and Environmental Impact Statement, Soil and Water Resources Conservation Act, Washington, DC: USDA.

Walker, R. 1982. T by 2000--Illinois erosion control goals. Land and Water, Urbana-Champaign, IL: University of Illinois Cooperative Extension Service.

CHAPTER 26

SHARING INFORMATION ON CONSERVATION TILLAGE

James E. Lake
Conservation Tillage Information Center
Fort Wayne, Indiana 46815

INTRODUCTION

A new Center for the collection and dissemination
of information on conservation tillage farming techniques
began operation in January of 1983. The establishment
of the Center, called the Conservation Tillage Information
Center (CTIC), culminated nearly 18 months of effort by
an ad hoc group repesenting the agricultural industry,
government agencies, private foundations, professional
organizations, and the National Association of Conservation
Districts (NACD).

The vision to establish the Center resulted from
the recognition that there is need for more exchange of
information between the private and public sectors, while
acknowledging the fact that the farmers' demands for sound,
up-to-date information on conservation tillage are increasing
daily. The Center's goal is to help fill these needs
by serving as a clearinghouse to increase the flow of
information on conservation tillage among agricultural
leaders in both the public and private sectors so that
it is more available to help farmers (and those agencies,
institutions, organizations, and industries that assist
them) to make informed decisions about conservation tillage.

It is no secret that America's farmers are currently
under severe economic pressure. As a result, they are
searching for new techniques to reduce input costs while
maintaining productivity. At the same time, their general
concern for soil conservation and water quality improvement
is on the increase. Thus, the demand for information

on conservation tillage has increased, accompanied by increased adoption of its use by America's farmers. Many industries, agencies, organizations and individuals are deeply involved in this change in agricultural technology. Prior to the establishment of the Center, there was little opportunity for exchange of information by these diverse groups.

CENTER IS A COOPERATIVE VENTURE

The Center operates as a cooperative venture between a number of private agricultural industries, the U.S. Department of Agriculture (USDA), the U.S. Environmental Protection Agency (EPA), and several non-profit organizations including: the Soil Conservation Society of America, the Iowa National Heritage Foundation, the Joyce Foundation, and the National Association of Conservation Districts (NACD). Many local conservation districts and farmer practitioners also have memberships in the Center.

OPERATION OF THE CENTER

Administration

The Center is being administered by NACD as a Special Project under the direction of an Executive Committee made up of agri-business representatives, professional society representatives, and farmers. An Advisory Committee for the Center, consisting of representatives of federal agencies contributing to the Center as well as the National Association of State Soil Conservation Agency Administrators and the National Association of County Agricultural Agents, has been established. The Center is headquartered in NACD's Washington, DC, office with a Field Office established in Fort Wayne, Indiana. Service support to the Center is also provided by NACD's five Regional Offices and its Service Department in League City, Texas. NACD's nearly 3,000 Conservation District Members and the 50 State Soil Conservation Agencies are providing vital outreach for the Center to gather and share information.

Agency Support

The Soil Conservation Service (SCS) - USDA, has provided a full-time employee to the Center. The SCS employee is located in the Ft. Wayne Field Office and serves as Field Specialist to assist with carrying out services of the Center. In addition, the agency is assisting in information sharing activities at the federal, state, and local levels.

The Federal Extension Service provided funds to cover 50 percent of the cost of employing a computer specialist from Purdue University for two months in the Spring of 1983 to help establish the Center's computer capability. In September, 1983, an Extension Specialist began a one-year detail assignment in the Center's Washington office. The Federal Extension Service is covering the salary and fringe benefits for the employee, while the Center is picking up travel and per diem expenses. We hope to extend this arrangement. State and local Cooperative Extension Service employees are also providing information to the Center and serving as resource specialists.

The Great Lakes National Program office of the U.S. Environmental Protection Agency is providing nearly $35,000 annually in direct financial support to the Center. Additionally, EPA has granted nearly $2 million to 33 conservation districts in Indiana, Ohio, Michigan, and New York to carry out conservation tillage demonstration projects.

Private Foundation Support

In August, 1982, the Joyce Foundation in Chicago provided a $50,000 challenge grant to NACD to help with the "start-up" costs of establishing the Center. The grant was provided on the condition that the agricultural industry would provide at least $25,000 in additional matching funds. That matching challenge has been met. An additional $25,000 was received from the Joyce Foundation in 1983. In December of 1983, the Shell Foundation provided a $50,000 contribution to help support the Center's operation.

The Iowa Heritage Foundation has provided the Center in-kind support in the form of staff time from one of their employees to assist in our membership drive.

Private Industry Support

Over twenty private industries representing equipment, chemical, seed and others, have already joined the Center as Corporate Members by providing financial support. The Dow Chemical Company provided the Center with a special contribution of $50,000 over and above their corporate membership to help cover the cost of employing a full-time Center Director who was hired in December. In addition to their corporate memberships, many companies like Chevron Chemical Company, DuPont, and Pioneer Hi-Bred International, Inc., are providing in-kind support to help promote the Center. Pioneer also released a major study on conservation through the Center for distribution which significantly enhanced public awareness of the Center.

SERVICES OF THE CENTER

The Center is facilitating the exchange of information through a number of services. They include:

Monthly Newsletter

A key element of the Center's information-sharing effort is the monthly newsletter called <u>Conservation Tillage News</u>. Each monthly newsletter contains the following feature sections:

<u>Research Watch</u>. Highlights on the latest research available on conservation tillage.

<u>Product News</u>. Introducing new and unique products and equipment for conservation tillage.

<u>Action Front</u>. Short articles on activities happening at the local level to share information on conservation tillage.

<u>Publications</u>. Highlights new publications on conservation tillage, along with details on how to obtain them.

<u>On the Calendar</u>. Provides information on upcoming meetings, tours, field days, exhibits, etc.

Each subject that is addressed in the newsletter is followed by a name, address, and phone number for the reader to do additional follow-up with the information source, if desired.

Resource Specialists Referral Service

The Center has established a list of Resource Specialists in each state. These specialists come from SCS, Extension, Agriculture Research Service, private industry, and other sources.

The Center sent each person a form asking him or her to catalogue their respective areas of expertise, be it weed control, insect control, disease control, soils, equipment, economics, etc. We then asked each person to sign the form indicating their willingness to serve as a resource person for the Center. All of these verified specialists are now on our computer file. The Center refers questions received from the person making the inquiry to the resource people from that respective area. The Center has established this service in full recognition of the site specific nature of conservation tillage. Information provided to farmers on conservation must be specific to the conditions of his or her area. The specialists serving as resource people are being asked to provide research information and field experience to the Center for use in the newsletter and other information sharing efforts. Participation by these specialists is one of the Center's primary sources of knowledge. Every person serving on the Center's "Resource People" reference list receives a complimentary newsletter. The Center is not and will not become a Center of expertise but, rather, a Center of information exchange--putting people with questions in touch with people with answers.

Computerized Abstract Library

The Center does not maintain a supply of publications on conservation tillage but instead is establishing a computerized abstract library to assist inquirers in locating sources of publications. We currently have over 500 abstracts of publications dealing with various aspects of conservation tillage. These abstracts have been developed by reviewing publications submitted to the Center by the Cooperative Extension Service, Agricultural Experiment Stations, Agricultural Research Service, Soil Conservation Service, EPA, private industry, and others. The computer program is set up so that we can search the abstract library using key words provided to us by the inquirer. The key words are used to search out only those publications which are appropriate to the inquirer's area of interest.

Audio-Visual Materials

The Center will utilize the facilities of NACD's Film Service Library in League City, Texas, to make slide tapes, movies and other audio-visual materials on conservation tillage obtained by the Center available to the public on a rental basis. The Center's newsletter will announce new audio-visuals provided to the film library, as well as other sources of audio-visual materials. Periodically, a catalog of the audio-visual materials on conservation tillage will be published by the NACD.

Annual Conservation Tillage Acreage Survey

Each year the Center will publish a comprehensive report on the number of acres of conservation tillage in the United States. The Center's first survey was conducted early in 1982 and published in June, 1983. Information for the 1982 survey was collected by the Center from SCS, Extension, and the State Conservation Agency in each state.

The 1982 Survey Report contains a listing of the conservation tillage in each state by conservation tillage type and by crop as defined by the Center. In addition, national and regional totals are presented along with various rankings of these states by conservation tillage types and crops.

In 1983, the Soil Conservation Service agreed to take the leadership in collecting the conservation tillage acreage information. SCS collected the information from each SCS Field Office on a county by county basis. Both SCS and the Center asked the Cooperative Extension Service, Agricultural Stabilization and Conservation Service and other agency representatives at the local level to assist the SCS District Conservationists in preparing the estimates for their respective counties. In addition, NACD asked conservation district officials and State Soil Conservation Agency representatives to assist the effort. The Center also encouraged local equipment, seed, chemical and other industry representatives to participate in preparing the estimates locally.

The Center received the county level acreage data from SCS in early December and loaded it all into the Center's computer to be analyzed and reported. The 1983 report contains more break outs of the data than the 1982 report. In addition, county level data (by state) will be available as supplements to the report. We also hope to do custom summaries upon request. For the 1983 report, the Center used definitions for conservation tillage established by SCS. The SCS definitions are very similar to the 1982 CTIC definitions with the only difference being that in 1983, the minimum residue requirements will be expressed as percent of cover on the soil surface rather than percent of previous crop residue. The reason for adopting SCS's technical specifications for conservation tillage is to establish a standard or baseline by which conservation tillage adoption can be evaluated in the future. The Center initiated the effort in 1982 to establish standard definitions for reporting all conservation tillage acreage and now SCS has helped by establishing technical specifications that the agency will use nationwide.

Opinion Survey on Conservation Tillage

In February, 1983, the Center mailed an "Opinion Survey On Conservation Tillage" to all conservation districts who agreed to participate with the Center. We asked them to seek local input in completing the survey. The survey contained questions such as: "What factors, if any, in your county are limiting the adoption of conservation tillage? For those farmers that have adopted conservation

tillage, why did they do so?"   The Center received nearly
1,000 completed survey responses.   The responses were
loaded into our computer at Ft. Wayne and analyzed.   We
published a report on this information last month entitled,
"Conservation Tillage: Local Actions and Views."   We
believe it will provide some very interesting insights.

Information on Demonstration Projects

Many conservation districts throughout the country
are currently conducting conservation tillage demonstration
projects.   In most cases, conservation districts have
made arrangements to provide conservation tillage equipment
for farmers to try conservation tillage on small acreages
of their land for comparison with their conventional tillage
system.   Results of demonstration projects of this type
are being reported through the Center.

Particular attention is being given to the conservation
tillage demonstration projects underway in the western
drainage basin of Lake Erie.   The EPA's Great Lakes National
Program Office has, over the last couple of years, funded
31 conservation districts in Indiana, Ohio, and Michigan
to carry out demonstration projects.   These 31 districts
blanket the entire western drainage basin of Lake Erie.
The districts are providing conservation tillage equipment
to participating farmers.   In addition, local SCS and
Extension personnel provide technical advice.   Local equipment,
seed, fertilizer, and chemical representatives are also
heavily involved.

For these projects, the conservation districts are
keeping detailed field records on all the inputs to the
conservation tillage plots as well as the conventional
comparison plots being conducted by their cooperating
farmers.   As part of an agreement between EPA's Great
Lakes National Program Office and NACD, the Center is
receiving all this field data on a plot by plot basis
for analysis.   A computer program has been developed to
input all the field data from the counties so that evaluations
can be conducted on such things as yields by tillage type,
limiting factors, amount of pesticides used, etc.   Results
of this comprehensive study are being published through
the Center.   The Center distributed over 1,500 copies
of last year's report.   We will have a report on the 1983
results out in late March or April.

FINANCING THE CENTER'S OPERATIONS

The Center is relying on membership dues, agency
support, foundation grants, and donations to finance its

operation. No regular conservation district dues to NACD are being used to help finance the Center. In establishing the Conservation Tillage Information Center, the Executive Committee established the following membership categories and fees.

Corporate Membership

Applies to commercial agri-businesses including equipment, chemical, seed, fertilizer, and others. Fees for Corporate members range from $500 to $5,000 per year depending on the volume of annual sales by the company.

Government Memberships

Agencies of federal, state, and local governments can join the Center by contributing skilled personnel or funds.

Conservation districts which agree to participate in the Center, assist in its survey gathering activities and provide a list of local resource people are automatically members of the Center. Resource people whose names have been submitted to the Center and who in turn agree to serve as resource information specialists for the Center are automatically members of the Center.

Institutional/Service Organization Memberships

Associations and institutions set up under non-profit charters, environmental societies, political organizations, foundations and other non-profit organizations - $100/year.

Resellers/Agricultural Service Firm Memberships

Wholesalers, retailers, dealers, banks, farm management companies, agricultural consultants, libraries, and other similar firms - $30/year.

Individual Memberships

Farmers and other self-employed individuals wishing to receive the Center's newsletter and be maintained on the mailing list - $15/year.

PARTICIPATION AND COOPERATION KEY TO SUCCESS OF CENTER

The Center has been established at a time when agricultural technology is changing rapidly. The success of the Center depends directly upon the willingness of key organizations, industries, agencies, and individuals to become involved in the information sharing activities of the Center. The Center relies heavily upon participation by local conservation district officials, SCS field representatives, local Extension Agents, university specialists, researchers, agri-business representatives, and farmers who are currently using conservation tillage systems as its sources of knowledge. With the continued support by these individuals, organizations, agencies, and industries, the Center will continue to be a valuable service to American agriculture and, at the same time, enhance the goals of conservation districts and NACD, those goals being the wise use of our soil and water resources.

SUMMARY

A new center for the collection and dissemination of information on conservation tillage farming techniques began operating in January of 1983. The Center is being administered by the National Association of Conservation Districts as a special project, and is a cooperative venture between a number of private agricultural industries, the U.S. Department of Agriculture, and the U.S. Environmental Protection Agency. Also involved in the venture are several non-profit organizations, the Nation's Conservation Districts and farmers interested in conservation tillage information. The goal of the Center is to increase the effective use of conservation tillage on America's cropland. The purpose of the Center is to increase the flow of information to agricultural leaders at the local level in both the public and private sectors in order that they might better serve farmers and others who request their assistance on a daily basis. The Center acts as a clearinghouse to put those people with questions on conservation tillage in touch with experts with answers. The Center recognizes the site specific nature of conservation tillage, and, therefore, does not attempt to serve as a center of expertise but rather as a referral service for the inquirers.

To accomplish its information sharing objectives, the Center carries out a number of services including (1) a monthly newsletter containing current information on conservation tillage. Key sections of the newsletter include: "Research Watch," "Action Front," "Publications," and "On the Calendar." Each subject addressed in the newsletter is followed by a name, address, and phone number for the reader to make contact for additional follow-up if desired. (2) A Resource Specialist Referral Service. The Center has a computerized list of verified State Resource Specialists who serve as Resource People for the Center. Inquiries received by the Center are referred to these specialists based on their expertise and areas of interest. (2) An Abstract Library. The Center has established a computerized Abstract Library Referral Service for helping inquirers locate sources of publications on conservation tillage. (3) Audio-visual materials. The Center utilizes NACD's Film Service Library to make slide-tapes, movies, and other audio-visual materials on conservation tillage available on a rental basis. (4) An Annual Conservation Tillage Acreage Survey. Each year the Center conducts a comprehensive survey on the acreage under conservation tillage throughout the United States. A comprehensive report of the survey results is published annually. (5) Assistance on conservation tillage demonstration projects. The Center provides assistance to Conservation Districts and others in setting up and carrying out demonstration projects on conservation tillage.